The Irish
Landscape

A SCENERY TO CELEBRATE

The Irish Landscape

A SCENERY TO CELEBRATE

Charles Hepworth Holland

DUNEDIN ACADEMIC PRESS • EDINBURGH

Published by
Dunedin Academic Press Ltd
Hudson House
8 Albany Street
Edinburgh EH1 3QB
Scotland

ISBN 1 903765 20 X

British Library Cataloguing in Publication Data
A catalogue record for this book is available from the British Library

Designed and typeset by Patty Rennie Production, Portsoy
Printed in Great Britain by CPI Group

Frontispiece: Lower Lough Erne evening.

Contents

PART 3: THE NORTH

PART 4: THE GEOLOGICAL BACKGROUND

Preface

This book is about the scenery of the whole of the island of Ireland in all its splendid variety. It relates this to the rocks upon which its nature depends. It is concerned with the history of these rocks, which culminates in what we now see and enjoy.

There are various ways to use the book. It can be read through from beginning to end. The short introduction to Irish scenery in Part 1 can be followed by Part 4 on the geological background, and then the detailed accounts of the scenery in parts 2 and 3. If a particular area is of interest that can be reached through the list of contents or the index of place names. Some necessary technical terms are given their own index, which relates to Part 4 and to any definitive material in Parts 2 and 3.

Many topographical maps of Ireland are available. The *Discovery Series* at a scale of 1:50,000 published by the Irish Ordnance Survey is to be recommended, as is the equivalent series for Northern Ireland. Useful geological maps of Ireland are the 1:750,000 sheet published by the Irish Ordnance Survey, based upon the work of the Geological Survey of Ireland; and the 1:250,000 map of Northern Ireland published by the British Ordnance Survey, but extending south to the 54th parallel and incorporating work by the Geological Surveys of both Northern Ireland and the Republic, together with that of W.S. Pitcher in Donegal.

What I have written depends upon the work of far more people than it is possible to acknowledge individually: colleagues, research students, friends, those many research workers who have written about

Irish geology and scenery. I thank the following in Dublin for their advice, for commenting on some parts of the text, for their encouragement: Declan Burke, Peter Coxon, Elaine Cullen, John Graham, Eileen Holland, Neil Kearney, Joann Layng, Corinna Lonergan, Edward McParland, Matthew Parkes, Tony Roche, Ian Sanders, George Sevastopulo, and Patrick Wyse Jackson. Sources of photographic illustrations are given at page 172. Elaine Cullen has skilfully drafted the diagrams. Declan Burke has applied his photographic expertise. I thank Patty Rennie for all the trouble she has taken in the design and typesetting of this book. I am very grateful to Douglas Grant and Anthony Kinahan of Dunedin Academic Press for their continued support in this enterprise and for all their kindness to me.

Fig. 1: Map of Irish counties.

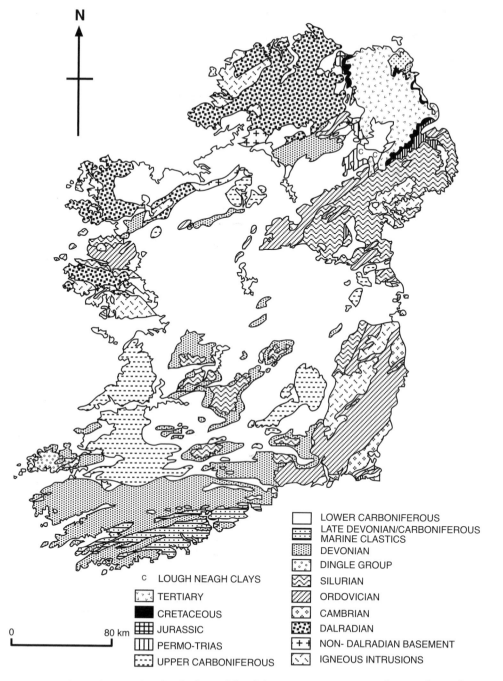

N

0   80 km

c LOUGH NEAGH CLAYS

TERTIARY

CRETACEOUS

JURASSIC

PERMO-TRIAS

UPPER CARBONIFEROUS

LOWER CARBONIFEROUS

LATE DEVONIAN/CARBONIFEROUS
MARINE CLASTICS

DEVONIAN

DINGLE GROUP

SILURIAN

ORDOVICIAN

CAMBRIAN

DALRADIAN

NON- DALRADIAN BASEMENT

IGNEOUS INTRUSIONS

Fig. 2: Geological map of Ireland; the width of the Cretaceous outcrop here and in other maps is exaggerated for clarity (compiled from many sources).

Eras	Periods			Approximate age in millions of years (Ma)
Cenozoic			Quaternary	
	Tertiary		Pliocene	1.8
				5.3
			Miocene	
				24
			Oligocene	
				34
			Eocene	
				55
			Palaeocene	
				65
Mesozoic			Cretaceous	
				142
			Jurassic	
				206
			Triassic	
				248
Palaeozoic		Upper	Permian	
				290
			Carboniferous	
				354
			Devonian	
				417
		Lower	Silurian	
				443
			Ordovician	
				495
			Cambrian	
				545
Pre-Cambrian eras			Proterozoic	
				2.500
			Archaean	
			Origin of the earth	4.600

Fig. 3: Geological time-scale.

Part 1

A SCENERY TO CELEBRATE

A Scenery to Celebrate

Sometimes on fine summers evenings we would leave Kruger's place and take the old mountain road north-eastwards above Dunquin to watch the sunset beyond the Blasket Islands, the 'ultimate shore' of the Old World. The narrow road itself, the scattered houses, the green pattern of the small fields are all human influences on this West Dingle landscape; but our eyes were for the more distant view: the last light on the shimmering sea, the brooding dark shape of the Great Blasket, the scattering of smaller islands, the sun dropping into the Atlantic beyond the islands.

To know the physical foundations of this landscape, does not diminish our appreciation, rather it increases it. Man, in his brief occupancy of Ireland, has done much to modify what we see in urban centres, in rural Ireland, even in what can still be called wild Ireland. But as we travel through the varied and often very beautiful landscapes of this island the very basis of what we see is provided by the rocks beneath our feet – and the story is not static; millions of years have led to this moment.

To glance briefly first of all at how very much there is to explain, we can begin a journey, as many travellers do, in Dublin. Here the River Liffey, which, with its 'hither-and-thithering waters', James Joyce used as a symbol of time, enters Dublin Bay, set between the high ground of Howth Head to the north and the hills of Dalkey and Killiney to the south. The river itself has a long history, as does sandy Bull Island within the bay. The so-called Leinster massif, which stretches from Dun Laoghaire and Dalkey south of Dublin to the Waterford coast has a

backbone of high mountains made of granite, with belts of older rocks on each side. Killiney Hill is made of granite. South of Enniskerry the Powerscourt Demesne is a popular visiting place. The Great Sugarloaf, its conical peak once wrongly referred to as a volcano, makes a splendid backdrop to the formal gardens. Within the mountains to the west and south are bleak peat covered moorlands, scattered plantations, and some attractive valleys such as Glencree and the Vale of Glendalough with its ancient stone tower and lake (Fig. 4).

South-eastern Ireland is a place of very varied and interesting geology, from the Kilkenny coalfield, to the intricate coastal morphology of the Wexford coast, to the very ancient rocks near Rosslare, and to the evidence of past volcanoes west of Waterford.

In a rapid clockwise circuit of the island, the interior must not be missed. To the north-west and north, through to central Ireland, mountain masses – the Knockmealdowns, the Comeraghs, the Galty Mountains, Slievenamon, Slieve Phelim-Silvermines-Devilsbit, Slieve Bernagh, the Arra Mountains, Slieve Aughty, and Slieve Bloom – which owe their elevation to the resistant Old Red Sandstone, stand above the lower ground of Carboniferous limestone. At the south-eastern end of the Comeragh mountains there is a rewarding ascent to the dramatic corrie of Counshingaun.

The Old Red Sandstone dominates County Cork, forming the highest mountains in all Ireland in Macgillycuddy's Reeks above Killarney. Popular since Victorian times, the Killarney area (Fig. 5) remains beautiful in spite of commercial distraction. In West Cork, an area of superb scenery, folding of the rocks and later erosion have resulted in higher ground of Old Red Sandstone alternating with valleys in the softer Carboniferous. A succession of peninsulas and attendant islands runs south-westwards into the Atlantic. A climb through the trees to the summit of the hill above the north-west corner of Lough Hyne (of which more later) to the north-east of Baltimore reveals a splendid view of this particular landscape.

Travelling farther north across County Cork into Kerry we come to the Vale of Tralee, which wraps around the dying slope of the Slieve Mish range at the landward end of the Dingle Peninsula. North of

Fig. 4: Glendalough, Wicklow Mountains National Park.

Fig. 5: Killarney National Park, County Kerry; Long Range and the Eagles Nest, looking towards the Upper Lake.

Tralee Bay, the Old Red Sandstone makes its last peninsula at Kerry Head. Now we are into Carboniferous country around the Shannon estuary and then all the way to Galway. The Shannon may seem to be small compared with such giants as the Amazon or one of the great Siberian rivers, but within Ireland it is long and substantial. Upstream from Killaloe it makes good country between Slieve Bernagh and the Arra Mountains.

There are two divisions of Carboniferous rocks in County Clare. To the south, the upper (younger) one provides spectacular coastal scenery from Loop Head to the famed Cliffs of Moher. Farther north is the special landscape of the Burren. The bare limestone surfaces of the Lower Carboniferous are crossed by joints enlarged by solution and rich in interesting plants. There are ephemeral lakes in depressions, caves below. The dry-stone walls themselves are of limestone. The same rock is seen also around Galway Bay and in the Aran Islands.

To the north is very different scenery. Connemara is a most beautiful place, depicted in many of Paul Henry's paintings. Irish scenery depends very much upon the ever-changing light. Here, once the rain stops and the mist clears, we see the striking shapes of the Twelve Bens (Fig. 6), mountains of quartzite standing above the blanket bog of their surroundings. This is an area of older rocks and a more complicated geology. There are different, more sweeping mountains in south Mayo, across the long narrow fjord of Killary Harbour. The Carboniferous limestone appears again around Clew Bay. The scattering of small islands are actually drumlins, glacial features here partly drowned by the sea. Inland they are to be seen across a whole belt of Central Ireland.

The mainland north of Achill and west of the Nephin Beg Range, together with the southerly directed peninsula of the Mullet, contains some of the oldest rocks in Ireland. There are desolate blanket bogs here concealing what is below, but with their own beauty in certain lights. On the bleak wind-blown western seaboard of the Mullet, sand-dunes, pushing their way eastwards, hide everything below.

The scenery of Western Ireland is so varied because of these numerous changes in rock type and in the ages of the rocks from place to

Fig. 6: Twelve Bens, Connemara, County Galway.

place. Farther to the north-east the Carboniferous limestones so typical of the Irish Midlands reach the coast again around Sligo and Donegal bays. Best known and outstanding in the scenery is the towering face of Ben Bulben. Sligo is closely associated with the name of W.B. Yeats. His poem 'Under Ben Bulben' anticipated his burial in the churchyard at Drumcliff, where his grandfather was Rector. Here again you may come upon the great house about which he wrote:

> *The light of evening, Lissadell,*
> *Great windows open to the south,*
> *Two girls in silk kimonos, both*
> *Beautiful, one a gazelle.*

9

In contrast, in the north-western corner of Ireland, in County Donegal, we are back to older metamorphic rocks and granites. Errigal, the highest and best known mountain in the county, is an example of the scattered quartzite peaks, recalling those of Connemara. The intricate Donegal coast, past Bloody Foreland and on to Malin Head, is exposed to the full force of the Atlantic and shows it.

Our clockwise progression leads next to north-eastern Ireland, to County Antrim and a very different scenery. Much younger rocks like those to be found in the south of England occur along the coast around the edge of the Antrim Plateau. Most conspicuous is the white chalk, harder than its equivalent in the white cliffs of Dover. The plateau itself is occupied by a wide spread of Tertiary basalt lavas. The central part of this has subsided to form the low-lying, clay filled Lough Neagh basin. The volcanic lavas are famously displayed on the north coast in the Giant's Causeway with its characteristic cooling columns (Fig. 7).

Farther south, the landscape from Longford, through Cavan, Monaghan, and Armagh, to County Down is obviously a geological extension of the Southern Uplands of Scotland. It continues the north-east to south-west grain of that country. These older rocks comprise coarser sandstone-like types and finer grained shales, the details of the rather subdued scenery responding accordingly. Younger granites cut through them in the Mourne Mountains. There are drumlin covered areas in the Longford Down belt where everything from drainage to road patterns to farm sites is controlled by the orientation of these mounds of glacial deposits. Around the Longford Down area, to the north-west, west, and south, the Carboniferous Limestone, though often concealed by glacial deposits, forms a large area from the Fermanagh lake country, through Roscommon, to Lough Corrib and Galway, and back eastwards through Athlone to the Dublin coast, where we began this journey.

We have had a rapid introductory tour through varied Irish landscapes. It has been difficult to avoid the use of a few geological terms. To look at the scenery in more detail, as later in this book, and to try to explain its origins, we need a wider vocabulary. The fourth part of

Fig. 7: Giant's Causeway, County Antrim; several individual Tertiary basalt lava flows are seen; a red fossil soil occurs near the top of the cliff.

the book is intended to provide this in a straightforward introduction to rocks and to the processes which affect them. A *selected index of technical terms* connects to this part and to explanations elsewhere in the text. There is also an *index of place names*.

Part 2

THE SOUTH

Fig. 8: Ireland's Eye, north of Howth Harbour.

Dublin and Howth

Holidays in Ireland often start, and finish, in Dublin. The capital city is well situated on a broadly curved bay south of the upstanding Howth peninsula, familiar to travellers arriving in Dublin by sea. To the south of the bay there is hilly ground from Dalkey to Killiney Hill, but the more conspicuous limit, corresponding in a way to Howth Head, is Bray Head, 8 kilometres farther south. The River Liffey enters the bay from the west; the Tolka to the north of it and the Dodder to the south make more modest contributions.

The oldest rocks in the Howth peninsula, as at Bray Head, are of Cambrian age. Referred to as the Bray Group, they comprise greyish, greenish, or red mudstones, siltstones, and coarser greywackes. Most striking, however, are the resistant light grey or yellowish quartzites, consisting of quartz grains cemented by silica, which form the summits of the peninsula and some conspicuous coastal features. Ireland's Eye (Fig. 8), the small island about one kilometre from Howth Harbour, has quartzites at each end and, in between, finer red or variegated rocks, partly covered at high water. The northern quartzite forms the main mass of the island. Its striking profile as seen from Howth was described by W.H. Bartlett in 1842: 'A huge rock on its eastern extremity appears to have been driven under by some convulsion of nature . . . .'

Though there has been additional evidence in recent years from microscopic marine fossils, the so called acritarchs (minute hollow organic bodies regarded as representing the reproductive or resting stages of marine algae), which can be extracted chemically from the

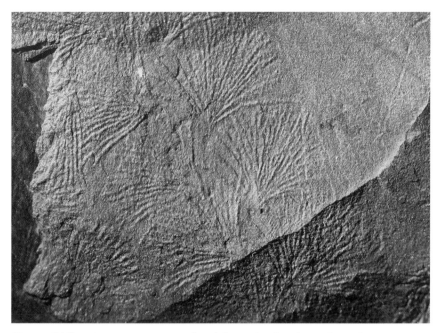

Fig. 9: Oldhamia antiqua, a trace fossil from the Cambrian Bray Group.

rocks, the Bray Group has traditionally been assigned to the Cambrian System on the presence of a fossil named in the 19th Century after a geological survey geologist as *Oldhamia* (Fig. 9). This is a *trace fossil*, which represents the activities of an animal, rather than the animal itself. *Oldhamia* can be found, but not too easily, in some of the finer grained, coloured rocks. It comprises a branching tunnelling system produced by some kind of worm. In three dimensions it can be seen to cut downwards through the original sedimentary layers. Elsewhere in the world it is associated with rocks otherwise known to be Cambrian in age.

The rocks of the Bray Group have been folded, cleaved, and faulted during the Caledonian earth movements (latest Silurian, early Devonian). Later Variscan (latest Carboniferous, early Permian) structures also are present. Furthermore it has been shown that some isolated bodies of quartzite, particularly those seen in the Howth Peninsula, are the result not only of the arrangement of tectonic folds but, in some cases, of earlier soft sediment deformation as masses of

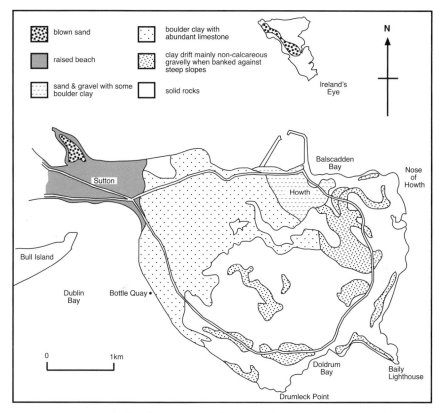

Fig. 10: Geological sketch map, Howth Peninsula, north side of Dublin Bay.

sand moved down the sea floor, perhaps becoming unstable because of local earth tremors. Other rocks may show these slump structures. Contorted strata at Sheep Hole on the south coast to the west-north-west of Drumleck Point, and Hippy Hole to the north-east of it, provide good examples.

The iron stained yellowish, brownish, and orange quartzite has long been quarried on a small scale at Howth to provide a building material seen well in some of the older walls (Fig. 11). It is also much used, but with less success, as a facing on some modern houses.

At Howth there is a large gap in age between that of the Bray Group and the next rocks to be seen. These are Carboniferous and thus some 200 million years younger. The relationship between the two systems is well seen in Balscadden Bay to the east of Howth village, where a

Fig. 11: Howth stone wall, Howth Peninsula.

steep zone of fractured and iron stained rock follows the line of faulting between the Bray Group and the locally brownish dolomitized
Carboniferous limestones and shales. It crosses the peninsula, though
perhaps not as a single fracture, to traverse the Sutton shore where
there is a beach-covered gap between the Bray Group and the Carboniferous. Beaches are, of course, ephemeral features.

The shallow water Carboniferous limestones and shales can be
examined in Balscadden Bay, where they contain marine animal fossils
such as the separated disc like ossicles of crinoid (sea lily) stems,
brachiopods (lamp shells), gastropods (sea snails), and the skeletons of
fossil corals. To the west of Howth harbour at the end of the sandy
beach different, not obviously bedded, limestones appear. It is instruct-
ive to throw a bucket of water over these to reveal the 'stromatactis'
structure, in which original spaces between living materials were filled
by normal sediment, whilst more enclosed cavities within algal fronds,
and layers in the skeletons of bryozoans ('moss animals'), sometimes
received a little gently deposited fine mud, but were later largely
replaced by growth of clear fibrous calcite, now seen as conspicuously

shaped masses within the solid rock These are the so-called Waulsortian limestones, named from a locality in Belgium. They are well
developed in Ireland. Farther west along this Claremont beach norm-
ally thin-bedded limestones appear again.

The rather featureless country through the city of Dublin and inland
is floored by Carboniferous rocks covered by superficial deposits, associated with glaciation and the rivers, but is interrupted in places by
small hills of fossiliferous calcite mudstone which developed as mounds
(isolated Waulsortian accumulations) on the original sea floor. There
are original depositional dips in such structures which slope outwards
from the centre. Feltrim Hill, (about 3 km south-west of Malahide),
now much quarried for road stone, is a familiar example often seen
from the air on the descent to Dublin Airport.

Variscan structures are represented by gentle folds in the
Carboniferous rocks and by faults, such as that separating these rocks
from the Bray Group. The occurrence of Bray Group rocks in Ireland's
Eye must be explained by upfolding or by the presence of a fault
between there and the Lower Carboniferous rocks of Howth. Beyond
this, a long period of time involving the latest part of the Palaeozoic
Era, the whole of the Mesozoic, and much of the Cenozoic is unrepre-
sented by obvious geological evidence in this area.

Local geological history is completed only by the rock record of the
Quaternary glaciation and post-glacial events, all to be fitted within the
geologically short period of about one million years. Evidence is to be
found in those superficial or drift deposits which cover much of the
surface (Fig. 10). From the melting ice itself came the boulder clay or
till, consisting of clay containing scattered and ill-sorted pebbles
carried within the ice. The pebble content of the till depends upon the
rocks over which the ice originally passed. For example, a calcareous
clay with debris from the Carboniferous is seen at the north-western
end of the Howth peninsula. A variety of less common pebbles and
boulders of igneous as well as sedimentary rocks may be identified in
places. Additional evidence of the passage of ice, and indeed its direction, may be seen in the form of striae on bare rock surfaces, where
they were scratched by fragments carried by the ice. On some lower

slopes there is much ill-sorted 'head' where material has been moved partly by ice and partly by late solifluxion (freeze-thaw) processes. Better bedded sands and gravels are found in places, as in the broad valley which runs towards the sea at Howth. These were deposited from water, trapped there by remnant ice.

The widespread melting of the European ice sheet some 10,000 years ago resulted in a rise in sea level. A subsequent relative elevation of the land surface has formed a raised beach now about 4 metres above mean sea level that is seen at various places along the Irish coastline. It forms the isthmus of sand and gravel now connecting the Howth peninsula to the mainland. Investigations during building operations and additional excavations made by the late Professor G.F. Mitchell revealed local details of the post-glacial period when Howth was still an island. A kitchen midden made by early Neolithic man was found here. Associated clays gave a radiocarbon date of 5250 ± 100 years BP (i.e. before 1950). At Bottle Quay, close to where Sutton Strand Road turns inland, material of the post-glacial raised beach rests upon a wave-planed rock platform which must itself pre-date the boulder clay which rests above it (Fig. 12). The boulder clay here is seen in two successive varieties, the lower containing shell fragments derived from the floor of the Irish Sea.

The environment is not static: geological processes of erosion, deposition, elevation, and depression continue. There is no better place to see this than in North Bull Island in Dublin Bay. This island of sand developed from a sand bank to the north of the line of the Liffey after construction of the South and North Bull Walls altered the configuration of the currents in the bay. Salt marsh encroaches more and more on to the sandy lagoon behind the northern portion of the island. A muddy area near the north causeway is bright green in summer and autumn with the succulent annual salt-loving plant *Salicornia*. Sand from the beach has been lifted by the wind to form a belt of dunes along the island, now somewhat stabilised by marram grass. Longitudinal ridges in the spread of dunes relate to earlier positions of the shore line. At the top of the sand beach, facing out into the bay, small barchans (crescent shaped sand dunes) such as one might see on a

Fig. 12: Raised beach, Bottle Quay, Sutton shore, Dublin Bay.

larger scale in a desert, may sometimes be observed to be moving along in the wind. The most prominent feature of the beach is an arrangement of ridges and runnels (channels) parallel to its length formed by wave action. The eastern end of the island is a curved spit bent by the refracted waves as it gradually grows. Since it first developed in 1902 there appears possibly to be a cycle of about 35 years before the recurve is destroyed by the waves and the cycle starts again.

South Dublin, Counties Wicklow, Kildare, Carlow, and Kilkenny

On the south side of Dublin Bay and south of the Carboniferous foundation of the city there is more variety in the geology. It is logical, and perhaps clearer as well as consistent, to begin with the oldest rocks. So first the Cambrian, which forms an elevated mass south-west of Bray Head as well as the Great Sugarloaf and Little Sugarloaf mountains. The outcrop continues southwards to the country west of Wicklow town. Comments on these Bray Group rocks already made for Howth apply again here. All these marine rocks were formed in a restricted, partly fault bounded basin which stretched north-eastwards across the area which is now the Irish Sea.

Inland from Bray Head, south-westerly striking quartzites contribute to the high ground. A walk around here shows various hollows and offsets of the quartzites relating to a series of cross-faults. The conical peak of the Great Sugarloaf (Fig. 13) forms an admirable back drop to the gardens of the Powerscourt Demesne. Its scree slopes of angular quartzite fragments are familiar to those who climb the mountain to take the splendid view across the city and Dublin Bay. The quartzites here dip generally eastwards as one limb of a syncline. This downfold, as with much of the Bray Group at Howth, was the result of original large scale soft sediment deformation. The later folds produced by the Caledonian earth movements are on a smaller scale and have an associated east-north-easterly cleavage. Along their north-western contact the Cambrian rocks were thrust to the north-west over the younger Ordovician. The small patch of Bray Group at Carrick-

Fig. 13: The Great Sugar Loaf, County Wicklow.

gollogan, clear even in the profile of the Dublin mountains seen from the north side of Dublin Bay (Fig. 14), is an isolated thrust mass.

The largest continuous area of granite in Ireland runs from the coast at Dun Laoghaire and Dalkey, south-south-westwards across County Wicklow and into Carlow and Kilkenny, forming the central belt of what is often called the Leinster massif. The old quarries at Dalkey provided granite for the construction of Dun Laoghaire harbour. Comments on this great intrusive mass come later; first we turn to the Ordovician. These rocks occur to the east of the granite on each side of the outcrop of the Bray Group and then extend south-south-westwards in a broad belt towards Enniscorthy. Another strip of Ordovician rocks lies to the west of the granite extending past Blessington.

In the eastern belt the lower part of the Ordovician is largely of deep water marine sedimentary rocks which become younger eastwards. Near the granite these rocks have been thermally metamorphosed into

Fig. 14: *View northwards from the Howth Peninsula, showing Dalkey Island, Killiney, and (behind) Bray Head, and the Great Sugar Loaf.*

schists. In the area to the south-east and down through Rathdrum and Avoca there are upper Ordovician slates and shales accompanied by volcanic rocks, many of them rhyolitic. These were associated with a continental plate margin below which at this time the deposits of the Iapetus ocean were being subducted (dragged down). There is good fossil evidence for the age of the upper Ordovician, with long known occurrences of graptolites (colonial animals distantly related to the vertebrates which secreted an organic skeleton), for instance in the railway cutting at Rathdrum, and other fossils shells such as trilobites (extinct arthropods) at Slieveroe.

The volcanic regime accounts for the mineral deposits long worked at Avoca. Copper, iron, sulphur, lead, silver, and zinc have been mined here intermittently since Prehistoric times. Some of the massive sulphides are seen to be intimately related to the stratification; probably they were exhalative deposits associated with the rise of metallic brines from the sea floor. There are also many old mining sites along the belt to the east of the granite and related to it. The old tips near Glendalough can still yield mineral specimens of galena or sphalerite, the lead and zinc sulphides, associated with quartz veins. And there is the long intriguing question of Wicklow alluvial gold in the Goldmines River near Woodenbridge. There are archaeological records; but the deposit was rediscovered in 1795 when 300 women with some men and children embarked on a veritable gold rush. It was quite successful. As recently as 1948, a diviner was used here to site three unsuccessful boreholes.

Beyond the other strip of Ordovician rocks to the west of the granite, the Silurian forms a roughly rectangular area. These rocks are greywackes and shales. One division with minor greywackes can be seen in a large disused slate quarry north of Blessington. The Silurian ground rises to the east of the Naas-Kildare road, which crosses the ill-exposed and low lying Carboniferous country, these rocks following unconformably upon the Silurian. To the west of the road a small narrow range including the Chair of Kildare marks the position of an interesting inlier of Ordovician, Silurian, and Old Red Sandstone rocks protruding through the Carboniferous.

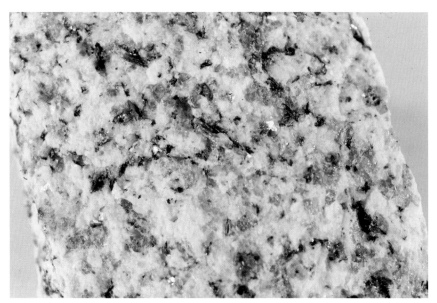

Fig. 15: A specimen of Leinster Granite showing quartz (glassy) feldspar (white), and mica (dark).

The Leinster Granite (Fig. 15), which was associated with the last phase of Caledonian folding, is of early Devonian age. Within its large extent there are several different phases of intrusion and varieties of granite. Thermal metamorphism by the granite has produced an aureole of schists, which can be seen above the beach at Killiney. There are slivers of schist along the spaces between different varieties of the granite; the Wicklow Gap crosses one of them. Although the granite forms upland areas they are generally rounded and covered by peat. It is better seen in the valleys, which tend to follow north-west to south-east trending fractures. More lush vegetation begins where the granite ends. The highest rugged ground tends to depend upon the presence of schists. The Lugnaquilla Pluton (Fig. 16) forms the most mountainous area but only because of resistant schist bodies within the granite. The highest peak at 925 metres has one such, which may represent part of its original roof. East of Carlow the granite is reduced to relatively low field covered ground of the Tullow lowland. Far to the east schists still form a higher rim, and farther south the Blackstairs mountains include isolated granite summits.

Fig. 16: Lugnaquilla, Wicklow Mountains.

The road through Carlow follows a narrow strip of Lower
Carboniferous rocks along the valley of the River Barrow. To the west
rises the rim of the Castlecomer plateau with its outcrop of Upper
Carboniferous strata of the small, once productive, Leinster coalfield,
which is almost a simple structural basin. The base of the slope is still
in limestone. Above are dark bedded cherts, a cryptocrystalline form
of silica. There follow grey sandy mudstones and pale sandstones with,
higher in the sequence, the so called Carlow Flags. Finally, on the
plateau itself which is something of a topographical basin, come the
characteristic cycles of beds of the Coal Measures: fireclays (fossil
soils), coal seams, freshwater and thin marine shales, and some sandstones. The environment was now one of swamps inundated at

intervals by fresh or marine waters. Associated with a thickening of the Jarrow Seam was a famous 19th Century find of fossil amphibia. Fossil plants are conspicuous in the Coal Measures.

Beyond all this, there is again a gap in evidence of geological history after the Carboniferous and before the glacial and periglacial events which have made such a difference to the scenery. The late Quaternary glaciation has left many physical features and deposits in this area. It probably reached its maximum about 20,000 years ago. There was a small dome of ice in the Wicklow mountains; ice flowed from the Irish Midlands; and a great sheet came down the Irish Sea. In the western part of the area most of the glacial and meltwater deposits came from the Midlands and thus Carboniferous limestone is conspicuous in them. The coastal strip from south Dublin to Wicklow has a clayey till with fragments of marine shells and erratics (foreign boulders and pebbles) from the Irish Sea. A walk on the beach at Killiney can produce an interesting collection. There are flints (cryptocrystalline silica again) from the Chalk and sometimes one is lucky enough to find pebbles of a distinctive microgranite, which comes from one small place in Scotland, Ailsa Crag in the Firth of Clyde. The traditional Scottish curling stones are obtained from it. Erratics from the Irish Sea show that the ice sheet, some 1000 metres thick, pushed up against the mountains, such as the Sugarloaf, to reach a height of several hundred metres.

Abundant water came from the ice. During the glaciation there was a glacial Lake Blessington, much larger than the present Pollaphuca reservoir, held against the western side of the mountains. Meltwaters coming from the Midland ice formed a deltaic complex. Substantial deposits of sand and gravel can be seen where these are being exploited (Fig. 17). They may be more than 80 metres thick and carry much lime-stone. There are large scale foreset and topset beds (respectively the coarser material dumped at the front of the delta and the finer mater-ial which flooded on top). Other lakes were held up where the meltwaters from the Wicklow ice cap were constrained by the margin of the Irish Sea ice. The Scalp, north of Enniskerry, may have been formed initially by subglacial drainage but is a good example of a melt-

Fig. 17: Deltaic foreset and topset beds, Blessington glacial lake complex.

water channel, as is the Glen of the Downs, inland from Greystone on the main road to Wicklow Town.

The glaciers expanding down the lower parts of pre-existing valleys over deepened them and moulded them into the characteristic U shape. Glenmacnass is a good example. The granite in the waterfall here is foliated (layered) and shows large aligned feldspars, related to the margin of the intrusion. We are here close to the metamorphic aureole in the schists. It is because of the resistance of the schist that such valleys plunge so steeply into more confined channels when they reach it. Another valley of characteristic shape and cross-section and other interesting features is Glendalough (Fig. 18) with its round tower and beautiful lakes. The delta here and the terraces show that the lake level was once much higher, perhaps held up by a moraine. Post-glacial detritus from a hanging valley has formed a fan separating the two lakes. A modern delta is growing at the top end of the second lake with detritus from the old mine workings.

Eventually small glaciers were left in corries where moraines now dam lakes, such as those of Lough Bray and Lough Nahanagan, the latter the site of a pumped storage scheme. When this lake basin was drained lake muds were found which gave a radiocarbon date of about

Fig. 18: Glendalough, County Wicklow; the two lakes are seen; Ordovician rocks in the cliff.

11,500 years. The moraine must have formed when there was a last cold burst after an interval of somewhat warmer conditions (the so-called Woodgrange Interstadial). In tundra conditions towards the end of this somewhat warmer period, the Giant Deer, with its huge antlers spanning as much as 4 metres, flourished briefly. The famous lake deposit at Ballybetagh Bog, north-west of Enniskerry, was a substantial source of their skeletons. It seems that the lake which once occupied the valley here was visited particularly by male deer and that unfit animals died and decomposed near the lake edge The antlers themselves, being fully grown, indicate autumn or winter conditions.

For the final Littletonian warm stage (in which we still exist) pollen has shown the changing vegetation of the woodlands and the time when eventually the first farmers took over. Blanket bog formed when conditions were wet enough.

The eastern limit of the area covered here is formed by the sea, where very different geological processes can be seen at work. The cliffs are constantly attacked as at Bray Head, and elsewhere beaches are built; longshore drift builds spits along the coast; coastal dunes are formed by the wind. The problems of maintenance of the railway track from Greystone to Wicklow alone testify to the dynamic conditions.

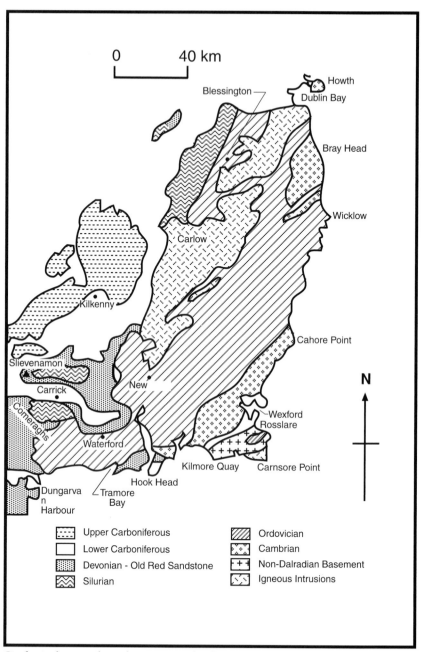

Geological map of south-eastern Ireland

Legend:

- Upper Carboniferous
- Lower Carboniferous
- Devonian - Old Red Sandstone
- Silurian
- Ordovician
- Cambrian
- Non-Dalradian Basement
- Igneous Intrusions

Map labels: Howth, Dublin Bay, Bray Head, Wicklow, Blessington, Carlow, Kilkenny, Cahore Point, Slievenamon, Carrick, New, Wexford, Rosslare, Comeraghs, Waterford, Kilmore Quay, Carnsore Point, Dungarvan Harbour, Tramore Bay, Hook Head

0 — 40 km

N

3

County Wexford

County Wexford is characterised by its fertile agricultural country but a remarkably varied geology is revealed in places. In the south-eastern corner of the county, isolated from other older rocks by an inconspicuous belt of Lower Carboniferous running north-eastwards to Rosslare Point and northwards across Wexford Harbour, is an area of difficult geology known as the Rosslare Complex. To the south-east, on the other side of this, at Carnsore Point, a granite is seen by the coastal walker; it is Caledonian in age, as are those in Wicklow. Another granite forms the Saltee Islands to the south of Kilmore Quay. The complex itself is of 'basement' rocks which have been affected by earth movements at various times since the late Precambrian. Small patches of Ordovician rocks overlie them. The Precambrian basement is of metamorphosed igneous and sedimentary rocks cut by north-north-easterly trending shear zones. Vivid testimony to the structural history of these rocks can be seen on the shore to the east of Kilmore Quay where the gneisses (themselves coarsely foliated) are seen to be tightly folded. The hinges of the folds plunge steeply downwards (Fig. 19).

Cambrian rocks, assignable still to the Bray Group, appear on the Wicklow coast near Cahore Point and continue south-westwards to reach the Wexford coast east of Bannow. Their most obvious contribution to the scenery is made by quartzites in Forth Mountain to the south-west of Wexford Town. A small strip of Cambrian rocks is seen again at the landward end of the Hook peninsula. At Booley Bay, at the western end of this, there is a splendid section of inverted shales and siltstones with well preserved sedimentary structures such as flute

Fig. 19: *Gneisses of the Rosslare Complex, east of Kilmore Quay, County Wexford.*

moulds. Technically these rocks are distal turbidites. The succeeding Ordovician with its volcanic components follows to the north, reaching the coast around Duncannon. We shall meet these rocks again in County Waterford.

The Booley Bay Formation is followed unconformably by Old Red Sandstone which passes transitionally into Lower Carboniferous calcareous rocks, which survive as the narrow rocky peninsula. The Carboniferous is exposed along the coast and at the headland itself, where the limestones are well known for their shelly fossils: corals, bryozoans, brachiopods, echinoderm plates, etc.

Apart from the older rocks, there are interesting geomorphological aspects to the Wexford scenery. Much evidence of the Pleistocene glaciation remains. For example, at Wood Village on the western side of the entrance to Bannow Bay the raised Courtmacsherry platform (which we shall encounter later) is cut into Cambrian rocks and overlain by beach deposits, 'head' (produced by solifluxion), and boulder clay.

The river system in south-eastern Ireland has long been the subject of discussion. The River Barrow, mentioned previously, runs southwards to reach the sea in Waterford Harbour. The Nore joins it from the north-west a little north of New Ross. Both these rivers, having their earlier courses in low lying limestone country on each side of the Kilkenny coalfield, have, farther south, cut deeply and discordantly across the older Leinster rocks. J.B. Jukes in a famous paper in the latter part of the nineteenth century suggested that the river drainage was superimposed from an original, but now lost, cover of younger rocks. Views are changing as the significance of uplift and other features of the Tertiary history of Ireland are becoming increasingly appreciated. The earlier effects of glacial meltwaters flowing south may also have been significant.

The Wexford coast shows many good examples of shoreline processes. Longshore drift is produced where dominant waves strike the shore obliquely. The ebb currents flow perpendicularly to the coast, so that material is gradually moved along as the returning waves come in. Sand and gravel spits may be built out across inlets in this way. This

Fig. 20: Tacumshin lagoon between Carnsore Point and Kilmore Quay, County Wexford.

effect is seen outside Wexford harbour. Along the south coast lagoons, such as that of Lady Island Lake and Tacumshin (Fig. 20) between Carnsore Point and Kilmore Quay, have formed where drainage from inland has been impeded by a bars of shingle covered by blown sand.

Before leaving County Wexford by crossing the River Barrow to Waterford City it is worthwhile to look at the rock relationships in the cliff high above the eastern side of the river and the station (Fig. 21). There is a clear angular unconformity–the closure of the Caledonian world one could say- between the gently inclined Old Red Sandstone and the steeply dipping grey Lower Palaeozoic rocks below.

Fig. 21: *Unconformity between gently inclined Old Red Sandstone and steeply dipping Lower Palaeozoic rocks, Waterford Harbour. Painting by G.V. Du Noyer of the Geological Survey (1866).*

4

Waterford and South Tipperary

The long outcrop of Ordovician rocks which sweeps south-westwards from Wicklow passes Waterford city and reaches the coast between Tramore and Stradbally. We are concerned here with an area which lay on the south-eastern margin of the ancient ocean of Iapetus, one of instability, earth tremors, volcanicity.

On the wave-cut platform south of Tramore, and through Lady Doneraile's Cove and Lady Elizabeth's Cove, dark grey and black, much contorted Tramore Shales are seen variously according to the state of the beach. Their age is probably early Ordovician. The Tramore Limestone of middle Ordovician age follows with some discontinuity. It is really a formation of calcareous shales and siltstones with limestone bands and nodules. From various places here and south-westwards, for instance at Newtown Cove, these rocks have yielded a rich shallow water shelly fossil fauna of trilobites, brach-iopods, and bryozoans. The fossils may be seen on water washed surfaces. The trilobites are particularly well seen in some inland expo-sures where the rocks have not been hardened by the effects of the sea. Westwards along the coast there are further faulted strips of Tramore Limestone and of dark shales, some of the latter containing graptolites.

But, above all, there are volcanic rocks, splendidly displayed. Sir Archibald Geikie, in his account of *Ancient Volcanoes* published in 1897, found here 'perhaps the most wonderful series of volcanic vents within the British Islands'. The volcanic rocks vary in age from place to place, and also in composition from basaltic through andesitic to rhyolitic. In the phase which postdated the deposition of the Tramore

Limetone acid tuffs were erupted explosively, some flowing down the sea floor. Rhyolites exploded from small vents, some of which were built up above sea level. These features can be examined in Lady's Cove and along Garrarus Strand to the west of it. At Lady's Cove the tuffs have been intruded by flow banded rhyolites and later by dykes of andesite. At Garrarus Strand the rocks include coarse agglomerates with blocks of rhyolite.

Inland in County Waterford and south Tipperary we begin to see a kind of scenery, and also changes in vegetation and fertility, dependent very much upon the rocks beneath. Lower ground of Carboniferous Limestone surrounds mountainous inliers of Old Red Sandstone, with still older cores of somewhat less resistant Ordovician and Silurian rocks. Farther west the Old Red Sandstone comes to dominate the whole scene. These arrangements are the results of folding followed by uplift and erosion. The major structures running approximately east-west involve Old Red Sandstone rocks of Upper Devonian age and also Lower Carboniferous, and thus must belong to the Variscan episode of earth movements of very late Carboniferous to early Permian times.

From the main road from Carrick-on-Suir to Dungarvan the impressive high ground of the Comeragh Mountains is seen to the west. It is good to climb the steep slope westwards to the formidable corrie of Coumshingaun (Fig. 22), cut into the steep eastern face of the somewhat flat-topped range. The back wall of the roughly semicircular corrie rises nearly 430 metres above its floor. There has been much past discussion about the origin of corries. They may have started with snow filled hollows. During the Quaternary glaciation, freezing water penetrated into cracks of the well-jointed and bedded rock to cause breaking off the face to maintain its steep profile. As is often the case, the corrie contains a lake, once dammed back at a larger size than now. Two terminal moraines are crossed as one approaches the corrie from below and there are remains of a lateral moraine to the south. Gullies are cut into the face of the corrie. Massive conglomerates of the Old Red Sandstone occupy the lower part of the walls and are succeeded by alternations of sandstones and conglomerates.

There are excellent views from the Comeragh Mountains (Fig. 23).

Fig. 22: Corrie in Upper Old Red Sandstone, Coumshingaun, Comeragh Mountains, County Waterford.

From north-east to south-east we can look over the Ordovician and Silurian inlier, which is really a low plateau 60 to 130 metres above sea level. It is cut by a narrow belt of Old Red Sandstone forming the low ridge of Croughaun Hill. The northern boundary of this is an unconformity and the southern one faulted. Solid rock of the Ordovician and Silurian is seen in old quarries, stream valleys, or road cuts. There are greywackes, siltstones, and purple and green slates.

The rocks strike west-south-westwards. This change of orientation from the prominent south-westerly Caledonoid trend in Leinster has developed through that second great period of earth movements, the Variscan, the effects of which became superimposed upon the Caledonian south-westward structural grain.

The Variscan folding itself is generally asymmetrical with steeper limbs to the north. Through this and original thinning there is but a narrow strip of wooded Old Red Sandstone to the south of the narrow and confined, Carboniferous floored, Suir valley with its town sites of Carrick-upon-Suir and Clonmel. Beyond this to the north there is a second inlier like that of the Comeraghs. Slievenamon itself is the Old Red Sandstone summit at the western end of it. Approaches from the north or south, as along the Kilkenny to Carrick-on-Suir road, involve climbs through the Old Red Sandstone to the so-called Ninemilehouse Tableland, a roughly rectangular area of Lower Palaeozoic grey banded and blue black slates seen in some old quarries, greywackes, siltstones, and mudstones. In this slightly more northerly inlier the beds strike south-westwards and show faults and cleavage which preceded Variscan events.

The few hundred metres of yellowish and greenish fine sandstones and shales at the top of the Old Red Sandstone around the Slievenamon inlier, south into the Comeraghs, and all the way across Ireland to the west coast in County Kerry, were recognised by the original geological surveyors as the 'Kiltorcan Beds'. There are large quarries above Ballyhale near the north-eastern corner of the Slievenamon inlier. The site of the original small old quarry at Kiltorcan, County Kilkenny, is close by. Here a rich fossil flora was first discovered in the middle of the 19th century. The plants include the large fronds of the fern-like *Archaeopteris* and various lycopods related to those which formed the great forests of later Carboniferous Coal Measure times. Fossil fish and the large freshwater mussel *Archanodon* were found with the plants. This must have been a special restricted lacustrine environment close to a forested area.

The generally anticlinal tract of Old Red Sandstone of the Comeraghs is continued westwards into the Knockmealdown Mountains. These are crossed by the road north of Lismore through 'The Gap'. North-westwards we look out across the Lower Carboniferous of the Mitchelstown valley to the Galty Mountains to be visited later. But next there is the extensive and splendid scenery of County Cork.

Fig. 23: *View from the north side of the Comeragh Mountains; fields on Lower Palaeozoic rocks, woods on Old Red Sandstone, Slievenamon in the distance.*

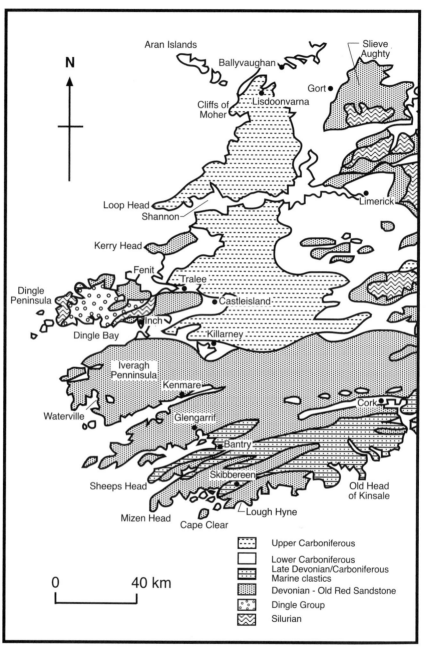

N

Aran Islands
Ballyvaughan
Slieve
Aughty
Gort ●
Cliffs of
Moher
Lisdoonvarna
Limerick ●
Loop Head
Shannon
Kerry Head
Fenit
Tralee
Castleisland
Dingle
Peninsula
Inch
Dingle Bay
Killarney
Iveragh
Penninsula
Kenmare
Waterville
Glengarrif
Cork ●
Bantry
Skibbereen
Sheeps Head
Old Head
of Kinsale
Mizen Head
Cape Clear
Lough Hyne

Upper Carboniferous
Lower Carboniferous
Late Devonian/Carboniferous
Marine clastics
Devonian - Old Red Sandstone
Dingle Group
Silurian

0   40 km

Geological map of south-western Ireland

5

County Cork and South Kerry

The geological map of Ireland shows an undulating line from Dungarvan, through Mallow and Killarney, to Dingle Bay. This is more or less expressed topographically as a boundary between the mountainous Old Red Sandstone country of County Cork and south County Kerry to the south, and the lower Carboniferous country to the north. It has been thought of in the past as one of structural change, a 'Variscan front', where more intense folding and well developed cleavage to the south give way to simpler arrangements farther north. It was sometimes thought of in terms of a thrust front with the implication of overriding from the south. In fact the tectonic style changes gradually northwards and, although there are thrusts near Mallow and some faults in other places, this is by no means everywhere the case.

To the south-west of Dungarvan there are several folds and cross-faults involving the Old Red Sandstone and Carboniferous. The limestone appears again at the coast near Ardmore, where the succession includes a quartz conglomerate. In places cleavage is pronounced. We have moved temporarily back into County Waterford.

Spenser wrote of the position of Cork: 'The spreading Lee that, like an island fayre, Encloseth Cork with his divided flood'. Robert Gibbings, in his book *Lovely is the Lee* published in 1945, referred to the two wide channels of the river reflecting the glittering limestone buildings. The whole position of the city – its old heart set on the limestone of the Cork syncline, its sandstone ridge rising to the north – reflects the Variscan structural grain. The white limestone buildings are still there; St Finbarr's cathedral, its interior with some interesting ornamental stones; the campanile of the church of St Ann with its bells of Shandon on the rising ground to the north; the older buildings of University College in its campus by the river. The limestones came from quarries in Little Island to the east of the city.

A good overview of the scenery and underlying geology of County Cork and south Count Kerry can be gained from a traverse starting at Killarney and running south through Kenmare and Glengariff, to Bantry. The town of Killarney itself is not distinguished, though its surroundings are simply beautiful (Fig. 5); they were already popular with tourists in the 19th century. The island scattered lakes on Lower Carboniferous ground are set against a half ring of Old Red Sandstone mountains to the west and south. In the west these rise to the summit of Carrauntoohil (1050 meters) in the massive range of MacGilly-cuddy's Reeks, the highest ground in all Ireland. The last Irish wild wolf was killed here in 1700. There are various west-south-westerly striking folds in the Old Red Sandstone of the whole Iveragh Peninsula. The Lower Carboniferous near Killarney shows a more complex detailed structure with minor folds, faults, and cleavage.

The scenery around Killarney owes much not only to the juxtaposition of Old Red Sandstone and Carboniferous but also to the effects of glaciation. U-shaped valleys, hanging valleys, corries, corrie lakes, ice scratched rock surfaces are all to be found. A good example of the corries is the Devil's Punch Bowl, high on Mangerton Mountain. Ice from a Cork- Kerry ice cap moved eastwards down the long valley now occupied by the Upper Lake and northwards towards what are now Muckross Lake and the Lower Lake, almost separated from it. The ice was sufficient to force its way also northwards between Purple Mountain and the Reeks, making the col of the Gap of Dunloe (Fig. 24). The summits of the Reeks evidently were never glaciated.

Contributing also to the Killarney scene, thanks to the mild climate, is the lush vegetation of the valleys and lower slopes, with natural oak woodland, the voraciously invading rhododendrons, the unusual *Arbutus* trees, and the parkland of the demesnes.

It will become clear that the Old Red Sandstone is not all of the same kind of rock. A total of some 3600 metres is present in this area.

Fig. 24: The Gap of Dunloe, Killarney.

Bright green massive cross-stratified sandstones, representing deposition in river channels, are seen in the Gap of Dunloe and near the Upper Lake. They are seen also as we take the Killarney-Kenmare road. There are lenses of conglomerate with quartz and reddish jasper (red chert) pebbles. Grey or purplish sandstones follow, the colour being due to a lower content of chlorite and the presence of ferric oxide. These rocks form the eastern part of the Reeks. The succeeding very thick purple sandstones are present along the northern foothills of the Reeks. A thinner development of greenish sandstones seen along the north shore of Muckross Lake allows separation from the highest formation of reddish purple, finer sandstones present only in the Muckross Demesne, before passage upwards into the so-called 'Lower Limestone Shales' at the base of the Carboniferous.

Moll's Gap, where the roads to Kenmare and Sneem separate, is another glacially created breach of a watershed. In the old sandstone quarries here fossil plants were found. The road continues downwards though green and then purple sandstones until the small town of

Kenmare is reached, situated where a bridge could be built across the narrow tidal reach of the River Rougty, before the widening but still narrow inlet of what is called the Kenmare River. The syncline here is tight, narrow, and complex in detailed structure. The Carboniferous rocks within remain on the land at each side only in limited areas towards the open sea. At Kilgarvan, some 10 kilometres up the valley from Kenmare, the thickness of the whole Old Red Sandstone succession has been estimated at over 6000 metres. All these rocks accumulated in the so-called Munster Basin. As they are all of continental (non marine) origin, this implies a subsidence of this magnitude. In places near Kenmare it is possible to see the rock record of the transition from continental Old Red Sandstone, through a tide influenced regime, into the marine conditions of the Carboniferous.

The road from Kenmare to Glengariff winds and tunnels across the wide wild tract of the Caha Mountains, passing as it does so from Kerry to Cork. There are fine views in both directions. It crosses also an anticline plunging westwards. The fine-grained rocks are a very monotonous purple and green sequence in which cleavage is well seen. There are small-scale ripple marks and desiccation cracks, testifying to the environment of a broad alluvial plain on the south side of the Old Red Sandstone continent, with its periodic flash floods and episodes of emergence. The glacially eroded Caha Mountains are scattered with lakes, such as Barley Lake to the west of the road.

Descending to Glengariff, once described as the most beautiful place in County Cork, we reach a coastal area containing the small harbour. It is rich in vegetation, including *Arbutus* and *Fuchsia*. The influence of the mild climate is seen at its greatest in Garinish Island with its Italianate garden protected by a perimeter of trees.

Around Bantry Bay we find a different scenery. By Devonian times the configuration of European geography had changed. The influence of Iapetus had now given way to that of a seaway farther south, from which there was marine transgression across the Old Red Sandstone continent. This reached what is now the far south of Ireland in latest Devonian times and then continued northwards during the Carboniferous. However, south of a line from Kenmare to Cork,

Carboniferous deposits in what has been called the South Munster Basin are not largely limestones as they are farther north. In the broad syncline which runs through Bantry Bay they are greyish sandstones which give way to a dominantly mudstone sequence and then to black slates. The last are seen also in Whiddy Island. They have yielded goniatites (earlier relatives of the ammonites) which indicate that the succession reaches the Namurian (lower part of the Upper Carboniferous).

For further exploration of Cork scenery it is convenient now to move to the south coast, whose general direction, in contrast to that in the west, is now parallel to the geological structures, with peninsulas perpendicular to it. A good example is the Old Head of Kinsale (Fig. 25) on the eastern side of Courtmacsherry Bay and some 50 km east-north-east of Skibbereen. At Holeopen Bay in the middle of the headland the sea has almost broken through. The Old Red Sandstone is not present here as it is in the Seven Heads, Galley Head, and Toe Head farther west. The uppermost Devonian rocks in the southern part of the peninsula are of greenish marine sandstones alternating with what are called heterolithic units; that is with bands of sandstone containing films of fine muddy material and mudstones with silt or sand lenses. Such deposits form at the present time, for instance on the tidal flats of the Dutch coast. In the narrow neck in the middle of the headland sandstones with larger scale cross-stratification appear, indicating higher energy tidal conditions. Higher still in the succession, where the peninsula broadens again northwards, there are grey mudstones with pyrite. Records of fossil spores have shown that the sequence here passes from Devonian to Carboniferous.

A characteristic feature of the southern coasts is a wave-cut platform or raised beach now lying several metres above the sea. This with the old cliff behind it is well seen in the neighbourhood of Courtmacsherry (Fig. 26). The many drowned valleys, such as the winding and narrow estuary of the Bandon River on the east side of the Old Head of Kinsale, represent sea level changes in the opposite sense, though the effect has not been sufficient to eliminate the raised rock platform which must be of pre-Glacial age.

Fig. 25: Old Head of Kinsale, County Cork; Devonian rocks in the foreground; Carboniferous beyond the narrow neck.

Lough Hyne, situated about 6 km south-south-west of Skibbereen and 5 km north-east of Baltimore, is a small but unique feature of Irish scenery (Fig. 27). It is a roughly rectangular marine basin with a longer north-west to south-east dimension of only about 1 km. Towards its western side the depth of the water reaches a maximum of nearly 50 m. Although the lough itself is relatively so deep, it is connected to the open sea though Barloge Creek by only a narrow and very shallow channel known as the Rapids. Here, over a sill, the water at low tide may be less then one metre deep. The tide flows inwards for about 4 hours and outwards into the sea for about 8.5 hours. In between there is a momentary slackness while the current changes direction. Van Gelderen gave a beautiful description of this: 'A hush falls over the waters, an unnatural quiet, and the seaweeds that have been pointing sea-ward quiver, stand up and, when the reversing current gathers momentum, fall back again into the water to flow now into the opposite direction, pointing inland. It is a moment when the sea and the lake seem to hold their breaths and if you were there once, sitting and waiting on the wall, you'd never forget it'.

The origin of Lough Hyne is something of a problem. Clearly marine erosion cannot provide the answer. Present day drainage to the lough is only by very small streams. The topography does not permit an origin from river erosion. As the surrounding rocks are Old Red Sandstone an origin by solution cannot be invoked. A tectonic origin has been suggested but there is no evidence for this. The Rosscarbery anticline simply passes across the basin; there is no significant faulting.

Fig. 26: The Courtmacsherry raised beach. Wave cut platform and old cliff.

The shape of the lough relates to the strike of the beds and to the joints which cross this approximately at right angles. There remains only one possible explanation: an origin by glacial erosion when sea level was much lower than at present. The well-jointed sandstones which cross the basin could have been plucked and scoured by ice moving from the north-north-east and meeting another stream from the east-north-east. The effect of this superimposition would be to produce the deep westerly trough. Jennifer S. Buzer investigated pollen and diatoms from cores through the sediment surface below the water. An original freshwater lake was affected by marine transgression about 4000 years ago.

Lough Hyne is a favourite haunt of biologists. It is rich in animal life, with fish, crustaceans, molluscs, and echinoderms. One has only to look down into the clear shallow water at the margin of the lake to see the scattering of deep purple spiny sea urchins on the bottom.

Having looked at Lough Hyne from close by it is worth the effort to climb the woodland path to emerge on the summit of Knockomagh (195 m) for a splendid view of the surrounding country). Out to the west are spread peninsulas and islands, largely of Old Red Sandstone but in places reaching into the kind of Carboniferous we saw in Bantry Bay. In Cape Clear the land is broken beyond Baltimore to produce Sherkin Island, Clear Island, and, out in the Atlantic, the well known Fastnet Rock. There are small islands in the bay to the north-west and then the next peninsula of Mizen Head. Beyond this again is Dunmanus Bay and then the very narrow peninsula of Sheep's Head. Beyond that again would be Bantry Bay. The pattern of Variscan anticlines and synclines along this coast is not so simple as it may appear. Many individual folds of various scales are involved; a well developed cleavage can frequently be seen approximately parallel to the axial planes of the folds. Examination of these rocks will show that the cleavage is more spaced out in the coarser rock types.

On Mount Gabriel (another good viewpoint) near Schull, on the way to Mizen Head, there are small mines where Neolithic man exploited the copper minerals from the Old Red Sandstone. Their method was to heat the walls of their small tunnels by fire and then to cool them rapidly with water. The rock thus shattered was extracted

Fig. 27: Lough Hyne, County Cork from Knockomagh.

with mining mawls, which are stones of suitable shape, some with a groove cut round them, so that a thong could be attached. The late J.S. Jackson described these and was instrumental in obtaining a radio carbon date of about 3220 BP from the charcoal remaining from the fires.

We now leave County Cork, but, before leaving also south County Kerry, there are several features to be mentioned in the broad Iveragh Peninsula of Old Red Sandstone, extending seawards from MacGilly-cuddy's Reeks. Here spectacular cliffs to the south of Dingle Bay, mountainous country including the summit of Coomacarea (775 m), and the lower rocky terrain along the Kenmare river contrast with the ill-exposed drift and bog covered country of the Inny Valley. A short way south from here, the small resort of Waterville is situated on a narrow moraine between Lough Currane and the sea. Inland the valley of the Cummeragh River leads to several mountain girt lakes of which the largest is Lough Derriaba; they are held by moraines.

Some 12 km beyond the end of the penisula small remnants of erosion of the highly jointed Old Red Sandstone form the Skellig rocks. Skellig Michael is a double peak reaching some 214 m high. An early Irish monastery was perched precariously here. Dr Michael Ryan has suggested that it may have been largely penitential.

The north-western corner of the Iveragh Peninsula has been separated by the sea to form Valencia Island. Fine-grained purplish rocks of the lowest part of the Old Red Sandstone here show a particularly well developed axial planar cleavage. They were widely used in the nineteenth century, the large slabs forming an admirable roofing and flooring material. They were used also in billiard tables.

In 1993 a most important palaeontological discovery was made by a Swiss geologist Iwan Stossel in the Old Red Sandstone on the north coast of Valencia Island. It is the meandering trackway of a primitive amphibian, itself probably about 1 metre in length, which wandered slowly across the Devonian alluvial flats, leaving about 200 footprints (Fig. 28). The locality is called Carraig na gCrub (Rock of the Hooves) which is about 4 km west-north-west of Knight's Town. Only about six such Devonian trackways are known throughout the world and the

Fig. 28: Footprints of a Devonian amphibian, Carraig na gCrub, Valencia Island.

Irish example is one of the oldest of these records of the first tetrapod animals to have evolved. The site is to be made accessible by footpath from the road, with a place to view the bedding plane with the footprints in no danger of damage to the site.

6

North Kerry

One way for us to travel on from Killarney would be to take the road north-westwards to Killorglin (site of the summer 'Puck Fair), which, like the river Laune which drains the Lower Lake, follows the outcrop of Carboniferous rocks towards the mudflats of Castlemaine Harbour at the head of Dingle Bay. The projecting Cromane Point which partly blocks the inner part of the silted up harbour is a moraine. Farther out towards the open bay there are two good examples of sand and shingle spits built out into the sea from near Glenbeige on the south side and, most impressively, from Inch to the north. Such features have been formed by longshore drift, though they were probably anchored originally by some configuration of the shoreline or quirk of the tides and currents. Their outer ends are hooked somewhat by the pattern of the refracting waves. Sand dunes add to their elevation.

A more directly northerly route from Killarney town is by the road to Tralee, through the glacial meltwater channel of Barry's Glen, now occupied by the River Gweestin, which crosses more elevated ground of Upper Carboniferous (Namurian) rocks. Some 200 to 300 metres west of where the road crosses the river, and north of Ballydeenlea, there is a small ill-exposed occurrence of Cretaceous Chalk forming the matrix of a breccia of Namurian shale fragments. This remarkable pocket within normal Upper Carboniferous rocks seems to have formed by a submarine collapse of the Cretaceous sea floor, probably by solution of Lower Carboniferous limestones at a lower level. In the past the chalk had been burned for lime, but in the 1960s P.T. Walsh brought it to the attention of the wider geological community, arranging for excavation and providing fossil evidence of age The locality is of great importance. Here are Upper Cretaceous rocks formed originally in the open sea and now at about 100 metres O.D. compared with the height of the Killarney mountains with their Devonian Old Red Sandstone at about 1000 metres. Evidence is accumulating of the importance of Tertiary (that is post-Cretaceous) history in the evolution of the Irish landscape.

The Dingle Peninsula as a unit is best approached as though travelling south-westwards by road from Limerick. The road crosses fertile Lower Carboniferous country until about 6 km beyond Newcastle West it suddenly bends and climbs the steep scarp forming the edge of a plateau area of Upper Carboniferous flaggy sandstones and shales. There is a stopping place by the road with a splendid view back. Some small hills made by the unusual development of volcanic rocks in the Lower Carboniferous can be seen to the south-east of Limerick city and there are areas of higher ground made by the Old Red Sandstone around and beyond. From the plateau there is an even more dramatic descent from its other side above Castleisland. The scarp sweeps concavely westwards above the fertile Vale of Tralee. To the west the anticline which brings up the Old Red Sandstone of the Slieve Mish range plunges eastwards into the vale. A dusty patch on the road to Tralee emphasises the presence of large limestone quarries. Beyond the city of Tralee the road runs north of Slieve Mish, following the Lower Carboniferous by the coast, with the ground rising very steeply to the south.

A steep climb on foot from the road through the wild country of Derrymore Glen cuts into the Old Red Sandstone to reveal in its higher part an inlier of fossiliferous coarse micaceous and calcareous Silurian rocks (part of what is called the Dunquin Group), upon which the Old Red Sandstone is unconformable (Fig. 29). Above to the south-west is the summit of Caherconree (827 m), the highest mountain in the Slieve Mish range. From the high ground here there are good views across Dingle Bay to the Iveragh Peninsula, with glimpses of the two spits referred to above.

A little before the small settlement of Camp a road forks right

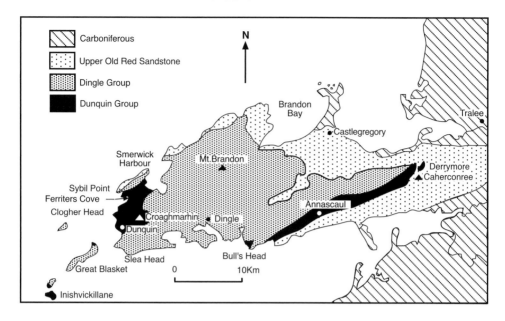

Fig. 29: Geological map of the Dingle Peninsula, County Kerry.

towards Castlegregory There are extensive beaches along here, then the peninsula which separates Tralee Bay and Brandon Bay, joined to the mainland by sand and sand dunes in which the small lake of Lough Gill is trapped. At the seaward end of the headland The Seven Hogs or Magharee Islands represent individual reef mounds like those we encountered in the Dublin region; the intervening rocks are beneath the sea. A road leads on past Brandon Bay through Cloghane but must stop before Brandon Point. West of here are formidable cliffs and above them the towering mass of Mount Brandon (953 m), Ireland's second highest mountain. The road directly from Camp leads south-westwards over the high Connor Pass to Dingle Town. The view north-eastwards from the summit is superb. Immediately below are small corrie lakes. It seems that the Kerry ice did not reach the summits of the Dingle mountains, but conditions were sufficient to allow corrie formation along the northern side of the Dingle Peninsula. Once on a cold moonlit night I left my car on the then rough road at the summit of the Connor Pass to take the view: the beauty was frightening.

The main road from Camp to Dingle passes through Annascaul. A little beyond Camp it crosses from the Upper Old Red Sandstone to rocks seen in no other area in Ireland: the so-called Dingle Group of late Silurian to early Devonian age. These continental (largely river borne) rocks include siltstones, sandstones, and conglomerates often purple in colour. Traces of the old Dingle railway are seen along this road. Soon a long straight valley appears to the south-east, formed by the softer often shaly or silty Silurian rocks of the Dunquin Group of the Annascaul Inlier. They are graptolitic in places, though, at the north-eastern end of the inlier, in faulted exposures of the kind of rocks we encountered in Derrymore Glen, and in others near Ballynane about 3 km east of Annascaul, there is a development with fossil shells. The inlier reaches the sea at Minard Head, where this strip has been compressed, crushed, and streaked apart. On the southern side of the small bay it is easy to see that there is faulting between these older Silurian rocks and the purplish Dingle Group which forms Minard Head. There is just a sliver of the latter. For most of the length of the Anascaul Inlier the Upper Old Red Sandstone is faulted against the shaly Dunquin Group; it forms a series of summits merging into the Slieve Mish range to the south-west of Cahereconree, though by then the Old Red Sandstone has become unconformable on the rocks below.

Having made a diversion from the main road to Dingle in order to reach Minard Bay it is worthwhile to examine some features along the coast of Dingle Bay. At Kilmurry about 3 km east of Minard Bay a minor road dips past the small ruin of Minard castle towards the sea, where there is a spectacular example of a storm beach with large smooth egg like boulders of Old Red Sandstone piled up by the most powerful storm waves and now stranded. The Old Red Sandstone along the coast is highly jointed into blocks which have been torn away by the sea, smoothed, and rounded at their vulnerable corners, and then heaped upon this beach. A little farther east the cliffs show an unusual development of Old Red Sandstone in that the sediments were not fluvial in origin but were deposited from the wind as sand dunes. The large-scale cross-stratification (Fig. 30) seen here is typical of dune bedding.

On the northern side of the Anascaul Inlier, as seen in Minard Bay,

Fig. 30: *Large-scale cross-stratification (dune bedding), Kilmuurry, north side of Dingle Bay.*

the contact with the Dingle Group is faulted. Farther west beyond Bull's Head there is an additional very small faulted inlier of the Dunquin Group and two small patches of Old Red Sandstone unconformable on the Dingle Group The locality is worth mentioning because Du Noyer, one of the Nineteenth Century staff of the Geological Survey, provided a striking illustration of the relationship (Fig. 31).

Beyond Dingle Town with its almost land locked harbour and bustling fishing port and tourist centre the road continues into West Dingle which has good inland and coastal scenery very obviously linked to the underlying solid geology. This is a relatively bleak area in the sense that trees are scarce and those present tend to be bent by the wind. In places a house will have its group of cordylines for, though there are high winds and Atlantic storms, the relative mildness is shown by the huge *Fuchsia* and, in summer the vivid orange patches of escaped *Crocosmia* (Montbretia) by the roadsides. A popular drive is round Slea Head and northward past Clogher Head and then inland through Ballyferriter. This reveals the scenery.

Fig. 31: Near Bull's Head, Dingle Peninsula. Old Red Sandstone unconformable on Dingle Group.

The road follows the coast along the southern flank of Mount Eagle and then precipitously round Slea Head, the Blasket Islands now coming into view. The road is cut in purplish rocks of the Dingle Group. They are seen dipping steeply southwards in the much photographed Dunmore Head. They make the main mass of the Great Blasket island, but its northern tip and the coast of the mainland northwards are formed by highly fossiliferous rocks of the Dunquin Group, faulted against the Dingle Group. Inland from Dunquin and south of the distinctive conical hill of Croaghmarhin there is stratigraphical continuity between the two groups.

The fossiliferous, more or less calcareous sandstones, siltstones, and mudstones of the Dunquin Group are accompanied by a remarkable development of volcanic rocks unusual within the Silurian of the British Isles. There are tuffs, agglomerates, lavas mostly rhyolitic , and rocks called ignimbrites which represent the rapidly cooled components of hot clouds of volcanic material, some of which have formed shards of volcanic glass. The rocks of the Dunquin Inlier are arranged in a substantial overfold (Fig. 32), its middle limb inverted. Variscan folds and fault and sporadic cleavage have been superimposed upon Caledonian structures. The volcanic rocks are thickest in the middle limb of the fold where they provide the backbone of the resistant mass of Clogher Head, as well as the good viewpoint of the inland crag of Minnaunmore rock to the east of the road. Silurian fossils (brachiopods, bryozoans, corals, molluscs, trilobites, etc.) are easily collected on the coast northwards from Clogher Head, on Drom Point (Fig. 33), and on to Ferriters Cove (Fig. 34).

Northwards from here the purple beds of the Dingle Group appear again in faulted relationship. The outcrop continues from Sybil Head across Smerwick Harbour. Beyond this a strip of Upper Old Red Sandstone follows unconformably upon the Dingle Group, dipping steeply north-westwards into the Atlantic. Out to sea this relationship can be glimpsed again in profile on the north-western side of the Blasket island of Innishtooskert.

The Dingle Peninsula is rich in archaeological sites, from the beehive huts seen between Ventry and Slea Head; to the little dry-stone,

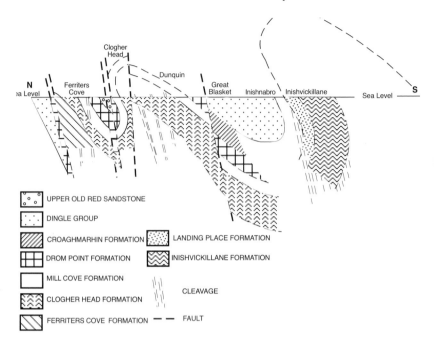

UPPER OLD RED SANDSTONE

DINGLE GROUP

CROAGHMARHIN FORMATION

DROM POINT FORMATION

MILL COVE FORMATION

CLOGHER HEAD FORMATION

FERRITERS COVE FORMATION

LANDING PLACE FORMATION

INISHVICKILLANE FORMATION

CLEAVAGE

FAULT

Fig. 32 (above): Diagrammatic section across the Dunquin inlier and Inish-vickillane, Dingle Peninsula.
Fig. 33 (right): The Dingle Peninsula, County Kerry. View northwards from Clogher Head to Drom Point; Mount Brandon in the clouds.

upturned boat shaped oratory of Gallarus (probably 12th Century) situated south-east of Smewick Harbour; to the various hill forts and promontory forts.

The Lower Carboniferous of the Vale of Tralee extends westwards along the north side of Tralee Bay as far as Fenit. Above the cliff about three kilometres north of here, on the north side of a small peninsula facing the end of a southwardly directed point, there are some most unusual karstic features in the limestone produced by solution, just possibly during Tertiary times (Fig. 35).

On the Atlantic coast farther north there is one remaining promi-nent peninsula of Old Red Sandstone, that of Kerry Head, which forms the southern margin of the mouth of the Shannon. The outcrop of Carboniferous rocks swings round the inland end of it to the east of Listowel, thus echoing the arrangement at the eastern end of the Slieve

Fig. 34: *View across Ferriter's Cove, Dingle Peninsula; the Atlantic cliffs are beyond the skyline*

Fig. 35: *Karstic feature in Carboniferous limestone, near Fenit, north of Tralee Bay.*

Mish range. In terms of past geography we are now outside the Munster Basin. The Old Red Sandstone here has a total thickness of about 700 m compared with the figure of over 6000 m near Kenmare. Sandstone bodies in the succession show much evidence of laterally migrating deep river channels. There are levee and flood plane deposits. Plant remains can be found in situ. The drainage was from the north.

7

Western Clare and southernmost County Galway

Across the Shannon estuary is County Clare. The western part of the county and its long coastline are in Carboniferous rocks; but there are two kinds: Upper Carboniferous (Namurian) to the south and Lower Carboniferous (Dinantian) to the north. These form two distinctive kinds of scenery which mingle somewhat in the middle.

The low lying Upper Carboniferous country north of the Shannon rises north-eastwards to form a bleak, wind swept, waterlogged, dissected, peat covered, sedge bearing plateau. The flat topped hills between Ennis and the coast reach a maximum height of about 390 m in Slievecallan. However, these rocks form some memorable coastal scenery.

To the south-west, bleak treeless country extends along the northern margin of the Shannon estuary to Loop Head with its lighthouse (Fig. 36). There are splendid views from here, southwards to Kerry and northwards to Galway. The general structure of all this Namurian country is relatively simple, with the beds not far from the horizontal, but there are local folds and faults. The formidable Atlantic seas have attacked the bedding and the joints to provide an excellent display of cliffs and sea stacks. There are sandstones here turbiditic in origin.

Elsewhere the lowest Namurian rocks below and partly equivalent to these sandstones are the dark Clare Shales which are widely developed. Changes in sea level allowed the presence of normal marine conditions at times. Above are largely mudstone slope deposits, showing the effects of slumping. A few kilometres north-west of Loop

Head, sand volcanoes have been described. Looking like small-scale true volcanoes but not more than 10 m in diameter, these are the result of expulsion of sand laden water from the surface of slumped masses (Fig. 37).

The highest Namurian rocks along the coast occur in a series of cycles from mudstone to siltstone to sandstone, representing the repeated incursions of deltas into the basin. Marine bands with goniatites occur at intervals and allow correlation of these rocks elsewhere in Ireland and across to Northern England. The best known and most spectacular scenery is at the Cliffs of Moher (Fig. 38), north of Liscannor Bay and west of Lahinch, in the northern part of the Namurian outcrop, where the upper part of the nearly horizontal sequence shows the lowest of these cycles. The cliffs are 8 km in length and nearly 200 m in height. The waves impinge directly upon them. Above the cliffs, part of the succession is of the Liscannor Flags, seen locally in field boundaries and widely used as ornamental stone in various places in Ireland. The meandering markings represent the trails of snails which originally crawled upon the Namurian basin floor.

The northern part of County Clare is the famed limestone area of the Burren. In limestone country the presence of carbon dioxide as well as water allows the slow solution of the calcium carbonate rocks and its removal as calcium bicarbonate Thus limestones make their own characteristic scenery (Fig. 39). The Burren, as is the case with west Yorkshire in England, has an additional special feature in the widespread development of limestone pavements.

The regional dip of the limestones southwards is very low, taking them underneath the Namurian rocks of south Clare. There are few slight folds as at Mullagh More on the eastern margin of the Burren. The ground itself rises northwards to reach a height of about 100 m above sea level overlooking Galway Bay.

The presence of the limestone pavements appears to depend upon an original cyclic pattern of sedimentation, probably related to changes in depth of the shallow Carboniferous shelf seas, producing subtle changes in lithology. The last component of a cycle is a uniform, massive, well-jointed rock which forms the surface of an individual

Fig. 36: Cliffs in Upper Carboniferous rocks, Loop Head, County Clare.

Fig. 37: Sand volcanoes in Upper Carboniferous rocks, on top of a slumped bed, County Clare coast.

pavement. Additionally, critical also was the glacial scouring of the rocks to remove any drift cover and thus to leave terraces of solid rock exposed to solution. Once the joint system is exposed at the surface, solution slowly widens and deepens the cracks to form the so-called grikes; the blocks of rock left standing between are the clints. Additionally again, there may of course have been a human element in the history of the landscape as it now appears.

With its particular rock composition and structure, its scarce tree cover largely confined to hazel scrub, and with its windy but moist and mild climate, the Burren is something of a botanist's paradise. There is a remarkable mixture of Alpine, Arctic, and Mediterranean plants thriving in the grikes and other small solution hollows. Maidenhair fern may fill the grikes. There are many flowers: a profusion of Bloody Cranesbill, gentians, orchids.

Characteristic of bare limestone country is the development of undergound drainage, of which there are many examples in the Burren, The cave system of Poulnagollum beneath the eastern side of Slieve Elva has been explored through 14 km of passages. Slieve Elva (344 m), to the north of the spa town of Lisdoonvarna, is a northern projection

Fig. 38: Cliffs of Moher.

of the Namurian cover, its waterlogged topography in great contrast to that of the near by limestones. Its surface drainage disappears in a series of swallow holes once the limestone is reached. Elsewhere on the limestone, dry valleys occur where an originally greater surface flow of water has disappeared underground, Such a valley leads down to the north coastal village of Ballyvaughan. The Caher River, which reaches

Fig. 39: Characteristic limestone scenery in the Burren.

the western coast south of Black Head, is exceptional in retaining its surface flow, its valley being choked with glacial drift.

Another characteristic feature of the Burren is the development of surface depressions on various scales, in many cases probably related to cavern collapse at depth. Lough Aleenaun, for example, about 6 km east-north-east of Kilfenora, which occupies such a depression, may fill to a depth of several metres in wet weather but dry out completely in the absence of rain. Springs enter it through joints and the water escapes through various swallow holes at the sides. Such self-contained ephemeral turloughs, as they are called, can be recognised, even when the water has gone, by the presence of a brownish black moss covering boulders. The turloughs are a rich source of Burren flowers. Some 6 km farther to the north-east is the Carran depression, the largest of all with an area of more than 7 km^2.

There are many ways to explore the Burren. A good starting point is Lisdoonvarna, itself on the Clare Shales of the Namurian. The road from here north-eastwards to Ballyvaughan is soon on to the limestone but there is an interruption in the small Namurian outlier of

Poulacapple. Soon after this the road falls steeply at Corkscrew Hill from which there are good views northwards.

The western coastline should not be neglected. There are splendid example in and above the sea cliffs of bedding planes and joint systems (Fig. 40). The coast from Crab Island north-eastward can be reached from Doolin. For a kilometre or so there are cliffs up to 20 m high, exposed directly to the full force of the waves (Fig. 41).

Various summit areas on the limestone ground of the Burren provide good views of the terrane. The top of Slievecarran in the north-east allows additional views across the Gort lowlands to the east and south. The small town of Gort itself is in a southerly projecting part of County Galway. In the area around Lough Cutra to the south-east there is a remarkable scattering of lakes which change in depth and

Fig. 40: Bedding and jointing in the Carboniferous limestone, Burren coast.

Fig. 41: The sea attacks the Burren coast.

shape as the supply of water varies. There is a complicated underground drainage system. Drainage from the high Old Red Sandstone
ground of Slieve Aughty farther east disappears into the limestone.
Variscan earth movements, responsible for the Slieve Aughty inlier,
resulted in relative elevation so that the limestone of the lowlands was
exposed to solution and erosion long before the Burren lost its
Namurian cover.

The Aran Islands which trend north-eastwards across the entrance
to Galway Bay belong geographically to County Galway but in term
of geology and scenery are obviously part of the Burren. They form a
ridge of limestone with impressive cliffs on the south- western side. The
limestone pavements here are bare and treeless. Small fields were
created by building dry-stone limestone walls and filling in with sand
and seaweed. Granite erratics on the surface were carried south by ice
from Connemara. The great stone fort of Dun Aengus above the

vertical sea cliffs provides one of many archaeological remains. John Millington Synge (1871–1909) spent the last decade of his life in the islands and drew his writings from them. Robert Flaherty's classic film *Man of Aran* was made here in 1934.

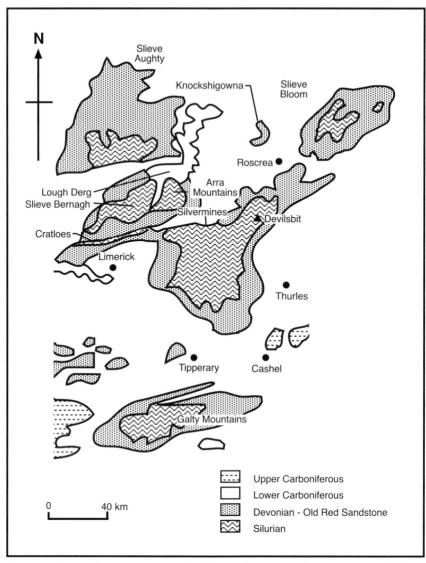

Fig. 42: Map of the central Irish inliers of Silurian (largely) and Old Red Sandstone rocks.

8

The South Midlands

In our progression southwards from Dublin, across to the west, and then northwards as far as Galway Bay we have neglected the heart of the South Midlands of Ireland. Here is ground of Carboniferous rocks much covered by glacial deposits, but there is an approximately V-shaped group of elevated areas formed by inliers of Lower Palaeozoic rocks followed unconformably by resistant Old Red Sandstone (Fig. 42). They comprise Slieve Bloom, Slieve Phelim-Silvermines-Devilsbit, Cratloe Hills, Broadford Mountains-Slieve Bernagh-Arra Mountains, Slieve Aughty, and, within the V the very small inlier of Knock-shigowna Hill (Fig. 43). It is convenient to add the inlier of the Galty Mountains in the south, though those of Slievenamon, and the Comeraghs have already been considered. Although they are basically similar, each of the central inliers has its own characteristics. Adopting a clockwise circuit again, we can begin with Slieve Bloom.

Travelling from Port Laoise to Mountrath to Roscrea we are conscious of a prolonged much forested mountain mass to the north-west. The Slieve Bloom mountains stretch for more than 24 km from Roscrea to Rosenalis in the north-east. Their maximum height is only 524 metres but their abrupt rise from the Carboniferois rocks of the plain make them seem something more dramatic. From Montrath roads lead to the mountain road which crosses the inlier to Kinnitty. On the top, where the road crosses the broad backbone of Slieve Bloom, as from the highest summit of Arderin to the south-west, we seem to have Ireland beneath our feet, the central plain with its fields, trees, and bogland stretching far into the distance.

Fig. 43: *The small inlier of Knockshigowna Hill has both shelly (in sandy pockets within conglomerates) and graptolitic Silurian faunas. View from the hill, as from an island, of the central plain of Ireland.*

Broad Variscan folding which has inherited an orientation from the earlier Caledonian structures has eventually resulted in a roughly oval area of Old Red Sandstone, these rocks providing, as it were, the strength of the mountains. The Old Red Sandstone, in all these central inliers is relatively thin at less than about 300 metres in total. Thus the particular feature of Slieve Bloom is that these rocks have behaved as a kind of carapace, through which erosion has penetrated centrally in a most irregular way to reveal Silurian rocks beneath in a curiously lobate axial area and some sixteen additional small inliers. The Capard ridge of Old Red Sandstone to the south-east, very clear in the topography, has a rather separate elongate inlier of Silurian rocks on its south-eastern face. To the north-west of it is the secluded deep valley of the Barrow.

The Silurian rocks are all of Wenlock age (the second of the four series into which the Silurian System is divided). Grey greywackes, characteristically banded siltstones, and shales dominate the succession, with graptolite fossils in places. These rocks are much folded and commonly very steeply dipping. The Old Red Sandstone is mainly of yellowish to purple to red sandstones with subordinate conglomerate and red siltstone. Their river borne nature is evident from much cross-stratification.

There is but a narrow strip of Carboniferous rocks between Borris-in-Ossary and Roscrea. Across it to the south-west begins the extensive inlier of Slieve Phelim-Sivermines-Devilsbit. The area of Silurian rocks within the Old Red Sandstone rim is here relative large and forms a much dissected group of summits, some of which are up to four or five hundred metres in height. These rocks are all Wenlock in age, small upright folds varying in orientation simply carrying the same part of the succession across almost all of the inlier. The lithologies are as in Slieve Bloom, including greywackes and the same characteristically banded siltstones, which in places are a source of graptolites and allow the dating of the rocks. They are seen typically by the road near Hollyford. Some modern working quarries show the somewhat different appearance of fresh unweathered rock. They also reveal a surprising number of small faults which must occur widely under the soil and drift cover.

The Old Red Sandstone of this large inlier extends southwards to within about 8 kilometres of Tipperary Town, in its Golden Vale floored by Carboniferous rocks and glacial deposits. It is convenient here to interrupt our journey by deviating farther to the south. A few kilometres from Tipperary Town the prominent wooded ridge of Slievenamuck is a fault line scarp along an outcrop of Old Red Sandstone which projects east-north-eastwards from the main inlier of the Galty Mountains. The widening gap between the two forms the Glen of Aherlow. This is a small enclave of wistful beauty, its good soils echoing those of the Golden Vale. Quaternary ice blocked its eastern opening to create a lake, whose various stages are shown by small deltas.

The Silurian rocks of the Galty Mountains inlier are mainly grey to greenish greywackes, siltstones, and mudstones, followed by darker shales with Wenlock graptolites. These Silurian rocks form a relatively subdued topography compared with the magnificent range of the Galty Mountains themselves and the lesser hills to the south and east, all formed by Old Red Sandstone, which comprises both aeolian dune deposits and alluvial fans. The summit of Galtymore (920 m), poised between the influences of Irish Sea and Munster ice, was itself ice free.

If we retrace our steps northwards, the north-eastern end of the Silurian outcrop near Devilsbit has special interests. There is a small area of younger Silurian rocks which are very latest Wenlock in age. On Devilsbit Mountain there are two isolated caps of conglomeratic Old Red Sandstone above the unconformity, with a prominent, partly fault controlled, gap between them. This is the well known Devilsbit itself which provides such a characteristic profile from various directions (Figs 44, 45). Legend has it that the Devil in angry form bit off the rock and spat it out as the Rock of Cashel. Unfortunately for the story the latter is of Carboniferous Limestone. There are few fossils other than graptolites in the Silurian but, near here, small coalified fragments of some of the earliest known land plants have been found in the marine siltstones.

In the north-west of the whole inlier, Keeper Hill (695 m) seems to tower over the landscape. It has a small capping of Old Red Sandstone

Fig. 44: Devilsbit Mountain with its caps of conglomeratic Old Red Sandstone.

remaining from the original cover. The north-western margin of the inlier near Silvermines is marked by a fault bringing the Silurian or Old Red Sandstone against the Lower Carboniferous, a narrow strip of which intervenes before more Old Red Sandstone in the Arra Mountains to the north. The faulted area has long been associated with metal mining and there are deposits of barite.

The line of faulting continues westward across the Shannon to form the northern margin of a long narrow strip of Silurian rocks in the Cratloe Hills. The name, which has been used since the original geological survey in the nineteenth century, refers to a small settlement just north of the Limerick-Ennis road and to an ancient oak wood on the south-western slope of Woodcock Hill. The course of the fault, which appears to have only a small throw, is marked by a depression. To the south the faulted margin of the Silurian outcrop is interrupted by areas where the unconformity with the Old Red Sandstone is still seen. The Silurian and Old Red Sandstone form a prominent ridge north of the City of Limerick, well seen from the Ennis road. Along the ridge the position of the boundary between Silurian and Old Red Sandstone

Fig. 45: One of Du Noyer's drawings of Devilsbit Mountain, seen from the south.

varies from place to place. Thus the prominent landmark of Gallows Hill at the western end of the inlier is in Silurian rocks, but the summit of the whole ridge in Woodcock Hill is actually part of the southern outcrop of Old Red Sandstone. Conversely, the prominent valley of Glennagross is occupied by a lobe of Silurian, the boundary with the Old Red Sandstone here swinging far to the south. Farther east, the summit of the small road at Windy Gap leading down into this valley is in Silurian rocks. And so the intricate relationship between rocks and topography continues eastwards.

The Silurian of the Cratloe Hills is similar to that in the Slieve Phelim-Silvermines-Devilsbit inlier and again there is evidence for a Wenlock age. In addition, though the exposures may no longer be found, there are important past records of shelly fossils – corals, brachiopods, trilobjtes, etc. – occurring in isolated conglomerates or pebbly sandstones. It seems that these were introduced along channels from the south, which would be nearer to the southern margin of the whole Silurian trough of deposition.

North of the Cratloes, the outcrop of Old Red Sandstone continues; then surrounds the Broadford Mountains – Slieve Bernagh – Arra Mountains inlier which is cut by the southern end of Lough Derg and the River Shannon, which here occupies a narrow valley where the river is bridged at Killaloe. Near Tomgraney to the north-west there are small faulted outcrops of Ordovician and Silurian black shales, but elsewhere the Silurian is of the now familiar greywackes and banded siltstones with graptolites in places. In the Arra Mountains shelly fossils were recovered like those found in the Cratloe Hills. The major structure of the whole inlier is a syncline, its axis running west-south-westwards across Lough Derg. But there are many minor folds and faults. A strong cleavage is developed in some of the finer grained units and slates have been quarried in places. The Silurian greywacke sandstones have proved to be relatively resistant, thus forming some of the highest ground, reaching over 500 m north-west of Killaloe and over 450 m in the Arra Mountains.

A strip of Lower Carboniferous rocks, partly occupied by the western lobe of Lough Derg and by the Scarrif River, separates Slieve Bernagh from the Slieve Aughty inlier to the north. Slieve Aughty in counties Clare and Galway was mentioned in the previous chapter as forming the eastern backgound to the Gort Lowlands. It is a dissected area of roughly radial drainage with a series of rounded or flattish summits, higher in the west where they reach as much as 400 m. The Lower Palaeozoic (some Ordovician but mostly Silurian) rocks here are a continuation of those in the Longford Down massif, itself a continuation of the Southern Uplands of Scotland. These rocks are much folded and faulted greywacke sandstones, siltstones, and shales of relatively deep water origin, in which there are graptolites rather then shelly fossils. The Caledonoid north-east to south-west strike has been modified here by subsequent Variscan movements, to give a more east-westerly orientation. The inlier is cut by a number of strike faults, some of which were reactivated after the Carboniferous.

The much more obvious rocks are of Old Red Sandstone, which forms rugged, stony, ground with peat deposits and heather prominent in the vegetation. It contrasts markedly with the green grassy areas of the Lower Palaeozoic. The erosion of the unconformable cover of Old Red Sandstone has left a complex margin to the large southern Silurian inlier in which lies most of the length of Lough Graney. The Old Red Sandstone rocks and, of course, the unconformity itself are affected by relatively simple Variscan folding. Otherwise the unconformity responds in only a minor way to changes in relief in the surface on which the Old Red Sandstone was first deposited.

The most striking scenic feature of the Slieve Aughty mountains is the contrast between the Old Red Sandstone and the Lower Palaeozoic rocks, and especially the way it is possible to see in so many places the precise nature of the unconformity between the two. The contact is well seen in the area to the west of Lough Graney where it is exposed in cliffs and crags for a distance of more than 1.5 km south of Drumandoora. It occurs also near the summit of Maghere, and north-east of Scalpnagown, where there are ten small outliers. The basal deposits of the Old Red Sandstone vary from place to place. Clasts of Lower Palaeozoic rocks may be included. There are good examples of the so-called cornstones which contain many nodular bodies of calcium carbonate, which material may also blanket and penetrate the uncon-formity. This is like the caliche deposits of present day semi-arid regions where calcareous soils are slowly formed.

North of Slieve Bloom and Slieve Aughty the country of the South Midlands merges into the great central plain of Ireland. Very obvious on the geological map is an almost continuous corridor of Lower Carboniferous rocks from Galway Bay though Ballinasloe and Athlone to Dublin. It is the tract familiar to travellers by train or road from Dublin to Galway; its scenery is not emphatic. The largely hidden surface of the Carboniferous limestone must have suffered much Tertiary erosion and solution to be followed by the effects of glacia-tion.. A very watery landscape followed with impeded drainage and many lakes. Gradually vegetation accumulated and about 7000 years ago peat began to form and to rise to result in the widespread raised bogs of the Midlands. During the glaciation streams within or under the ice carried sand and gravel which remained as sinuous ridges when the ice had gone. These are the eskers well developed in this Midland stretch (Fig. 46). They provided ancient dry routes across the country and even firm beginnings for bridging points of the rivers, as across the Shannon at Athlone and Banagher.

The corridor of Carboniferous outcrop makes a clear and useful demarcation between the solid geology of the south and that of the

Fig. 46: Eskers, Ballinlough, County Roscommon.

north of Ireland (nothing political about this) (Figs 1, 2). To the north there are substantial areas of relatively older and, paradoxically, younger rocks; there is much variety; there is wild and beautiful scenery.

Part 3

THE NORTH

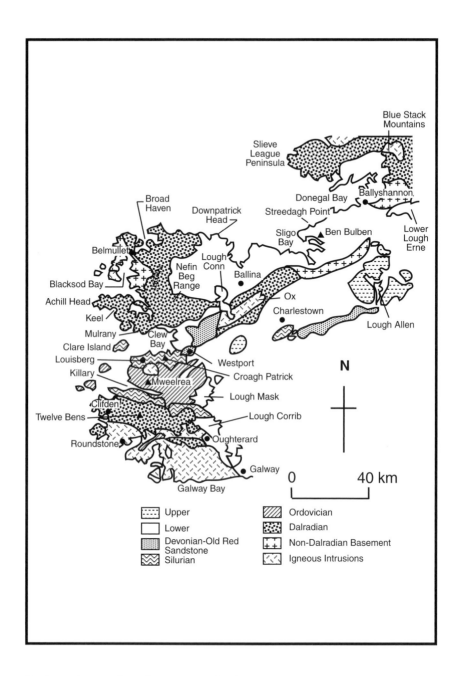

Geological map of north-western Ireland

South-west County Galway
(Connemara)

Connemara can best be taken as the area which lies north of Galway Bay, east of the Atlantic, west of Lough Corrib, and south of the unconformable cover of Silurian rocks in Joyce's Country. It comprises very largely rocks of the Dalradian Supergroup of late Precambrian age, which were metamorphosed in the Ordovician Period and intruded by granites in Silurian-Devonian times. A broader northern belt of the east-west trending Dalradian rocks is of schists, quartzites, and marbles representing an original sedimentary series. A southern belt somewhat south of the Oughterard-Clifden road is of gneisses resulting from the metamorphism of basic and other igneous rocks intruded into the Dalradian earlier than the main granites. We shall encounter Dalradian rocks also in North Mayo, Donegal, and Antrim. They are widespread in the southern part of the Scottish Highlands.

Glacial erosion along the Carboniferous limestone against the older rocks of Connemara must have been important in producing the deep basin of Lough Corrib (Fig. 47), which reaches below sea level. Its exit is over the salmon weir to the east of the cathedral in Galway City. Taking the coast road to the west through Spiddle we are on a large composite granite body intruded into the Dalradian, the Galway Granite itself. Inland to the north glacial erratics are scattered. The straight east to west coastline of the granite gives way in Bertraghboy Bay to intricate inlets and islands. In fact this granite is not very resistant; it does not form high ground as do many other Irish Caledonian granites. South-west of Roundstone a separate granite unit has been

Fig. 47: Lough Corrib evening.

cut into by the sea to form Gorteen Bay and Dog's Bay (Fig. 48), the two separated by white sand which is made of the small shells of Formainifera (protozoans). North of the small area of granite that remains here, altered basic and ultrabasic igneous rocks, and gneisses derived from them spread all the way from the east of Bertraghboy Bay to Slyne Head. This desolate area of blanket bog is scattered with very many small lakes. Blanket bog has formed in very wet climatic conditions on mountain slopes which are not too steep; in the west of Ireland it occurs additionally near sea level. The peat is thinner than in the raised bogs of the Irish Midlands. Stumps of pine trees are often found below the peat, indicating an earlier forested situation. The isolated summit of Errisbeg (300 m) provides an excellent viewpoint across this country and towards the mountainous area of the Twelve Bens. The road from Roundstone continues through Ballyconneely to Clifden, the main settlement of Connemara. It passes Mannin Bay where the beaches are often referred to as 'coral strands'. They are actually of fragments of the calcareous alga *Lithothamnion*, which, like the shells in Dog's Bay, have been swept in from the sea floor.

The principal route from Galway to Clifden leaves Oughterad close to Lough Corrib and runs westwards through Maam Cross and Recess. It leaves another granite near Lough Boffin and then follows the southern part of the schist and marble country of north Connemara. Between Lough Shindilla and Lough Oorid we have the first substantial mass of quartzite mountain to the north which leads northwestwards into the Maumturk Mountains. North of Recess, from the low quartzite summit of Lissoughter, the Twelve Bens also are brought into view. The valley containing Derryclare Lough and Lough Inagh curves round the eastern end of the Twelve Bens. The greenish Connemara Marble is present in places. Now sold largely as small ornaments it makes a fine ornamental stone, as seen for example in the Museum Building of Trinity College Dublin. The green colour results from the mineral serpentine, an hydrous magnesian silicate produced by silicification and metamorphism of original dolomite.

From many directions the quartzite peaks of the Twelve Bens dominate the scene (Fig. 49). The highest peak Benbain (730 m) is

Fig. 48: Dog's Bay, Connemara.

Fig. 49 (previous page): Twelve Bens, Connemara, across blanket bog.
Fig. 50 (above): Twelve Bens, Connemara evening.

surrounded radially by others. This is the very special scenery of Connemara seen in the changing light. The white quartzite mountains may stand out glistening in the sunshine or make dark silhouettes (Fig. 50); there is the desolate wet blanket bog of the valleys and surrounding areas; there are the colours of the lichens and the heaths.

The road from Clifden curves north-eastwards to Letterfrack, originally a Quaker settlement and now with its well placed information centre for the National Park. Diamond Hill, just to the east, is another good view point. The road continues eastwards past Kylemore Lough with good views again towards the Twelve Bens.

2

South Mayo and North Galway

The road from Kylemore Pass crosses Owenduff Bridge where, nearby, the river enters Lough Fee and continues to the shore of Killary Harbour. Leenane, near the eastern end of this long inlet, is a good centre from which to explore South Mayo, a beautiful area of wild mountains and coast. Joyce's Country is used for the area of mountains and valleys to the west of the isthmus between Lough Corrib and Lough Mask. The Kilbride Peninsula, within it, lies between Lough Kilbride and Derry Bay, both extending west-south-westwards from Lough Mask. A belt of folded and faulted Silurian rocks extends from here west-north-westwards along the south side of Killary Harbour. It is unconformable on the Dalradian of Connemara. Murrisk is the large roughly quadrangular mountainous area, north of Joyce's Country, which is bounded by Lough Mask and a Carboniferous outcrop to the east, by Clew Bay to the north, and by the Atlantic to the west. It is largely of Ordovician rocks disposed in a broad synclinal belt through the Partry Mountains, the Mweelrea Mountains, and the Sheefry Hills. To the north of this on the south side of Clew Bay is a structurally complex belt of Silurian and older rocks.

The Silurian of Joyce's Country begins in the Llandovery (the first of the four series into which the Silurian System is divided). Shelly fossils are known, for instance along and above the small road running though pleasing scenery on the south side of the Kilbride Peninsula. There are graptolites in places, for example at Doon Rock above the road. Higher Wenlock graptolites have been found elsewhere, as at Owenduff Bridge previously mentioned. The thick Silurian succession

Fig. 51: Killary Harbour.

as a whole begins with some river borne coarse purple beds; continues through a varied marine sedimentary succession dominated by turbiditic sandstones and mudstones; and finally becomes shallower, until at the top there are red beds which may be of very shallow water origin. These are seen for instance around Killary Bay Little.

The Ordovician rocks of what can be called the South Mayo trough have a total thickness of over 6 km. Killary Harbour, which is a glacially over deepened and subsequently drowned valley (Fig. 51), developed on relatively soft green slates. The Mweelra Mountains to the north are formed from a unit which makes up about a third of the whole thickness of the Ordovician. Coarse-grained cross-stratified sandstones representing a fluvial plain dominate the succession. Several extensive bands of ignimbrite punctuate the sequence. The road from Leenane which crosses the Erriff valley at Ashleigh waterfall and then turns south-westwards along the north side of Killary Harbour runs below a scarp of these sandstones. Two oblique streams on the south face of Ben Gorm mark the course of an important pair of north-westerly trending faults, the Maaam Faults, which shift the western part of the axis of the Murrisk Syncline by some 5 km to the north-west. Mweelrea itself, which is the highest mountain in Murrisk, reaches 827 m. The road to the west of Ben Gorm passes the small Fin Lough with the beautifully situated lodge named Delphi at its head. Doo Lough, the black lough, follows and the road then runs northwards to Louisburgh and Clew Bay. Lough Doo is situated on the same formation of Ordovician slates as exploited by Killary Harbour. Below these and farther down in the north limb of the great syncline are turbiditic light weathering sandstones set in mudstones with black laminae. The Sheeffry Hills are to the east and the lowest part of the Ordovician here is of similar rocks but with many pyroclastic levels. Both these last formations have yielded lower Ordovician graptolites. Like other finds in the Ordovician of South Mayo they relate to the north-western (Laurentian) margin of the Iapetus Ocean.

A belt of Silurian rocks extends from the Atlantic coast through the fine mountain of Croagh Patrick (Fig. 52), and then narrows until it reaches the Carboniferous unconformity to the south-east of Westport. To the west these rocks form lowish ground. The succession is metamorphosed to the extent that white mica and the green mineral chlorite have appeared There are sandstones, siltstones, and impure limestones of shallow water marine origin, but the most conspicuous rock is the quartzite which forms Croagh Patrick. Its ascent is easy if steep and the wide track of angular scree is familiar to the thousands of pilgrims who every year climb to the summit to be rewarded, if the mist clears, by one of the most splendid panoramas in the whole of Ireland. To the north stretches Clew Bay (Fig. 53), scattered with innumerable very small islands, most of which are partially drowned drumlins (Fig. 54). Inland to the east is Carboniferous low land and in the distance to the north the wild mountains of the Nephin Beg range.

An area of Silurian rocks forms the southern margin of Clew Bay from Roonah Quay eastwards through Louisburgh to Old Head. It is separated from the Croagh Patrick belt by the Emlagh thrust, which has had substantial lateral movement. There are many folds and faults. The rocks are cross-stratified sandstones and siltstones. In a thick

Fig. 52: Croagh Patrick, County Mayo, *from the west-north-west.*

Fig. 53: Clew Bay, County Mayo, looking northwards.

Fig. 54: A drumlin, Clew Bay.

middle unit there are interbedded mudstones which may be red in colour. Desiccation cracks have suggested a sand flat – playa lake complex. Higher siltstones have yielded non-marine fish and a curious frond shaped fossil which can be matched precisely in Silurian rocks of the Midland Valley of Scotland. The non-marine succession is completed by coarse sandstones occurring in channels, well seen at Roonah Quay. The same succession of Silurian rocks is seen in Clare Island, which guards the entrance to Clew Bay. Here the Silurian is unconformable on older rock.

A thin strip of these older rocks occurs also on the mainland along the south side of Westport Bay and eastwards between the Silurian and the Carboniferous. It is in fault contact here with the Silurian. This is a very complex belt of controversial origin which may relate to the extension of the Highland Border fault system in Scotland. Dated

Ordovician rocks are involved and north of these Dalradian or earlier rock. The latter includes an interesting metamorphic rock, green serpentinite, whose parent would have been ultrabasic. It may be encountered on the north-eastern flank of Croagh Patrick.

3

North Mayo

In south County Cork we saw that marine transgression northwards across the Old Red Sandstone continent was already beginning in very late Devonian times. The first Carboniferous marine beds appear later and later in the rock succession as we move northwards through Ireland. In the north the terrestrial influence tends to persist. On the north side of Clew Bay a Carboniferous non-marine unit produced by rivers flowing off the continent to the north is followed by some transitional beds, and only then by the fully marine Carboniferous limestones of the kind we have encountered so frequently so far.

The scattering of drumlins in Clew Bay extends inland around Westport town, concealing the Carboniferous there. To the north-east of Newport a valley leading past Beltra Lough as far as Lough Conn marks the prolongation of the Carboniferous outcrop. On each side of it in broken faulted patches are Old Red Sandstone rocks. Material flowed in various ways into a continental basin and the predominant rocks are conglomerates. They form Croaghmoyle mountain to the east of Beltra Lough. There are some impressive exposures of the conglomerates. Outside the two developments of Old Red Sandstone there are Dalradian rocks.

A partly fault bounded strip of Precambrian rocks extends north-eastwards from here through the Carboniferous lowlands. It crosses the south end of Lough Conn and Lough Cullen eventually to reach County Sligo. South of Sligo Bay it becomes a narrow prong concave north-westwards to die away near Manorhamilton. The whole feature can be referred to as the Ox Mountains though the topography is very varied, the mountainous ridge reaching over 500 metres in height beyond Lough Cullen, but becoming disconnected and glacially eroded in the north-east. The north-eastern part of the inlier is of gneisses which form the oldest part of the succession. Their age has proved to be controversial but they now appear to be equivalent to the oldest part of the Dalradian. They were derived originally from sedimentary rocks but are themselves intruded by igneous bodies which have also been metamorphosed. The more typical Dalradian rocks in the south-west are cut by a large body of granodiorite penetrated by the end of Lough Conn and its immediate junction with Lough Cullin at Pontoon.

The road west from Newport follows the narrow outcrop of Carboniferous rocks along the north coast of Clew Bay to Mulrany. The massive dissected Nephin Beg range with its connected and disconnected summits lies to the north of the road. The high relief is once again dependent upon the quartzites within the Dalradian outcrop which crosses the range. These form a roughly triangular area from a base stretching east-north-eastwards from near Mulrany to the fine peak of Nephin, which reaches just over 800 metres. There is much evidence of glaciation in the corries which cut the range, but the high summits seem to have stood above the ice. Pollen analysis from cores has shown that this bleak area of mountains was once forested.

From the south-east corner of Blacksod Bay the sea has exploited the relatively soft Dalradian schists in an intricate way, in the south almost cutting off the quartzite of Curraun Hill. In this sheltered area at Mulrany the vegetation is profuse; it includes the Mediterranean Heath. The road turns north-north-westwards and then south-westwards around the north side of the Curraun peninsula to cross to Achill Island at Achill Sound. This large island is all of Dalradian rocks, with the now familiar pattern of quartzite mountains and the rest of bog or heath on the softer schists or other metamorphic rocks. There is complex folding and faulting. As we take the road to Keel, Knockmore and Minaun rise to the left. The road continues to Dooagh. There are places where very small amethysts have long been

Fig. 55: Achill Head, County Mayo, by Paul Henry (1918–20), oil/panel-board, 23x31 cms.

collected from the residual sand. This is really a form of quartz in which a ferric iron impurity has produced the purple colour.

Finally we come to the spectacular north-western end of the island (Fig. 55). To the east the splendid conical Slievemore rises directly from the sea to reach a height of 670 metres. To the west, Croaghaun is almost as high. It faces into the relentless Atlantic which has produced one of the most formidable shear cliffs in all Ireland. The high ground here gives excellent views of the Mayo coast and inland to the south-east. There are small corrie lakes produced by a local ice cap. To the west a serrated ridge leads out to Achill Head.

The road northwards from Mulrany to Bangor Erris and thence to Belmullet passes through one of the greatest and most desolate extents of blanket bog in Ireland. It could be described as uninteresting, but its extent is impressive and the ever changing light can do things to Irish landscapes. To the west is the indented coastline of Blacksod Bay. The blanket bog is underlain largely by Dalradian rocks. At Belmullet is the very slight land connection with the curiously shaped Mullet peninsula. On the peninsula, the solid rocks are much obscured by bog and blown sand; but there are coastal exposures, especially along the Atlantic coast. Most of the outcrop is of the so called Annagh Gneiss Complex. There are two anticlines plunging to the south-east which take the complex across to a little exposed area on the mainland around Doulough. These gneisses are older than the Dalradian and have suffered pre-Dalradian deformation and metamorphism, as well as being involved later along with the Dalradian. They were formed originally from igneous rocks. Something of their complexity may be seen in cliff sections such as at Annagh and near Doolaugh. The west coast is exposed to the full force of the Atlantic and sand dunes push eastwards across the peninsula. There is evidence of several particular episodes of dune formation. Along the shore of Blacksod Bay low cliffs can be seen cut into the peat. At the far south end of the peninsula a boss of Caledonian granite penetrates the gneisses and now forms Temon Hill.

The area east of Broad Haven and north-west of an undulating line from Bangor to Port just to the north-east of Belderg on the North Mayo coast is occupied by Dalradian rocks. There is very good cliff scenery from Benwee Head eastwards to Glinsk, where the highest cliff occurs. East of this Dalradian area and of the Nephin Beg range, and north-west of the Ox Mountains, we enter an area of Carboniferous

Fig. 56: Doonbristy sea stack, Lower Carboniferous, North Mayo.

rocks centred on Ballina and Killala Bay. The coastal section east of Port demonstrates the continental influence on these Carboniferous rocks. Fluviatile deposits give way transitionally to coastal marginal marine rocks. At Downpatrick Head, for example, the sea stack of Doonbristy (Fig. 56) shows transitional deposits with a muddy channel, followed near its top by fully marine limestones and shale. Higher in the North Mayo succession there is a return to a mixture of fluviatile and marine conditions in the so-called Mullaghmore Sandstone Formation before the more typical marine limestones form the ground from Killala Bay through Ballina and to Lough Conn. The direct road from Bangor to Ballina at first climbs to elevated and desolate country in the Carboniferous continental rocks. Near Crossmolina the scene changes to fertile farmed country on the drift covered limestones.

4

Sligo, south-west Donegal, Fermanagh, Leitrim, Roscommon

A narrow strip of Carboniferous rocks runs north of the Ox Mountains, envelops Sligo Bay, expands towards Ballyshannon in south-west Donegal, and wraps around Donegal Bay. It expands again eastwards into Fermanagh and Leitrim and then southwards across Roscommon to reach Athlone in the north Midlands, where we left it in the previous part of this book. Similarly in treatment, it is convenient to take this large rather irregularly shaped area as a whole before we move northwards into the larger part of County Donegal.

Sligo Town is situated on the short river which drains Lough Gill into the sea. The lough itself is beautifully situated between the Ox Mountains and the area of mountain to the north; it contains the well known island of Innisfree. Neglecting a very small inlier near Rosses Point of metamorphic rocks like those in the north-eastern Ox Mountains, the whole area between Sligo and Ballyshannon is occupied by Carboniferous rocks The structure is relatively simple and the impressive scenery depends very much upon the various lithological contrasts.

The road from Sligo runs northwards through Drumcliff where, as mentioned in the first part of this book, W.B. Yates is buried. On the north side of Drumcliff Bay, Lissadell faces south. The coastal area through Grange is occupied by muddy limestones and shales. At Serpent Rock in the west and Streedagh Point, farther north-east along the coast, there are rock surfaces with remarkable displays of silici-fied fossil corals which thrived in the shallow marine waters of

Fig. 57: Streedagh Point. County Sligo. Lower Carboniferous corals.

Carboniferous times (Fig. 57). Some of the skeletons are bent, having collapsed and then grown upwards again towards the light. The occurrence was described by Wynne in 1864 as '. . . like stumps in a cabbage garden, and one is almost disappointed to find that one can not pull them up . . .' Some reach over half a metre in length. Farther along the coast, sandstones higher in the succession have been faulted into place; they form Mullaghmore Head, after which this formation is named. Along the coastal road there have been good views of the imposing face of Ben Bulben, which, though reaching over 500 metres above sea level, is not so much an isolated mountain as a scarp face (Fig. 58).

The uppermost part of the scarp face is of a massive, well jointed, limestone with many chert bands and nodules. Below are less pure limestones with shale bands which decrease upwards. The lower part of these and a formation of relatively soft shales below are covered in scree. It was the undercutting of the shales which led to collapse to a steep scarp. The Glencar glaciated valley to the south of Ben Bulben shows the landslips thus caused. The top of the Ben Bulben Range and

Fig. 58: Ben Bulben, County Sligo.

the Dartry Mountains to the north-east form a plateau with bog developing above the limestones and swallow holes through it. Farther back from the scarp, and particularly in the Dartry Mountains, patches of shale and a succeeding sandstone overlie the limestones.

North of Ballyshannon, after which the limestones mentioned at Serpent Rock and Streedagh Point are named, the Carboniferous outcrop widens into the Donegal Syncline. The rocks here show a transition from fluviatile to fully marine conditions, similar to the situation in North Mayo. Ballyshannon is on the river which drains Lower Lough Erne. The lough extends east-north-eastwards on the Lower Carboniferous towards the Omagh Syncline and then bends to the south-east. Like Lough Mask, it has a glacially deepened basin which falls below sea level. The River Erne past Enniskillen drains from Upper Lough Erne which has an extraordinarily complex arrangement of small islands and associated lakes. This is another example of a landscape formed by drumlins; it is part of a drumlin belt which extends across Fermanagh, Monaghan, and Armagh, to County Down.

Drumlins are streamlined mounds, usually of boulder clay, with their blunt ends formed upstream. They have clearly been moulded by moving ice, but their original distribution may relate to irregularities in the rock surface below or to some kind of stress pattern.

The succession on each side of Upper Lough Erne rises into the Upper Carboniferous (Namurian). The larger area of these rocks is in the west in counties Sligo, Leitrim, and Roscommon, where it forms high ground converging southwards on Lough Allen. Arigna, on the south-western side of Lough Allen, gave its name to a small coalfield where coal seams within the sandstones and shales were exploited for two centuries. The Cuilcagh Mountains in the north-eastern part of this area are significant in providing the source of the River Shannon which flows through Lough Allen, passes Carrick-on-Shannon, and descends gently southwards forming lakes in places down through Lough Ree (another irregularly shaped lake with some evidence of a contribution by limestone solution) to Athlone.

The large area of Lower Carboniferous rocks forming the North Midlands spreads eastwards from Lough Mask, across Roscommon and South Leitrim with their waterlogged or drumlin scattered landscape, across the Shannon, and eastwards again. In his novel published in 2002, *That They May Face the Rising Sun*, John McGahern evokes life around a lake, probably in Leitrim:

> The night and the lake had not the bright metallic beauty of the night Johnny had died: the shapes of the great trees were softer and brooded even deeper in their mysteries. The water was silent, except for the chattering of the wildfowl, the night air sweet with the scents of the ripening meadows, thyme and clover and meadowsweet, wild woodbine high in the whitethorns mixed with the scent of the wild mint crawling along the gravel on the edge of the water.

Extending west-south-westwards from near the southern end of Lough Allen a long narrow strip of country almost bisected by Lough Gara involves the Old Red Sandstone of the Curlew Mountains to the east and Lower Palaeozoic rocks to the south of Charlestown in the west. The Old Red Sandstone is faulted against the Carboniferous to the north and at both ends of its outcrop. The Carboniferous is un-conformable to the south. Massive conglomerates and associated cross-stratified pebbly sandstones represent alluvial fans and river deposits. Overlying are sandstone and mudstone flood deposits of an alluvial plain. The so-called Charlestown inlier, like the Old Red Sandstone, is not well exposed. A small area of steeply folded varied Ordovician volcanic rocks and black mudstones with early Ordovician graptolites is followed unconformably by Silurian rocks of early Silurian (late Llandovery) age. The first deposits are fluviatile in origin but the remainder, including a local limestone, is rich in marine fossils, especially corals.

5

Donegal

Donegal, a relatively large county, occupies the north-western corner of Ireland. It is an area of great beauty, even by Ireland's high standards. A glance at a general map of the solid geology may suggest relative simplicity. Dalradian rocks occupy the majority of the county from the Atlantic coast and extend eastwards outside it as far as the Sperrin Mountains in west Derry and north Tyrone. The only other conspicuous rocks are the Caledonian granites of western Donegal (Fig. 59). In fact, generations of geologists have worked towards an understanding of the relationships of these rocks and of their very complicated tectonic structures.

There is a general north-east to south-west grain to the country, though the main streams actually flow to the west or to the east from a watershed well to the west. They must have been initiated on some pre-glacial surface. During the Quaternary glaciation, of which there is much evidence, ice moved radially from a Donegal ice cap on this watershed, which was itself breached in places by the retreat of corries. The area is sliced by faults following the Caledonoid trend. Some are lines of shattered rock, others are thrusts, or folds developing into thrusts; still others have significant lateral movement. The Dalradian began as a varied and very thick sedimentary succession related to the north-west (Laurentian) margin of the Iapetus Ocean as that was being initiated. There were phases of earth movement and metamorphism, particularly in the early Ordovician, when these late Precambrian sandstones, mudstones, and limestones became the quartzites, schists, and marbles which we now see. Thus a stratigraphical succession, now

Fig. 59: Well jointed Donegal granite.

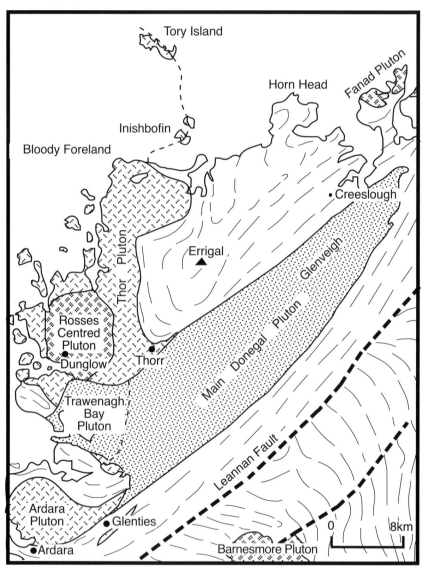

Map of Donegal with granites.

highly folded on various scales and tectonically interrupted, has gradually been clarified and related to the Dalradian of Scotland. Then there
are the Caledonian granites of different forms and relative, though
close, ages. There are also many minor basic intrusions.

Leaving the north shore of Donegal Bay westwards we pass the
fishing port of Killybegs and are into the Slieve League peninsula,
which supported its own small ice cap. There are several quartzite
bodies in the peninsula. Slievetooey, on the north coast, reaches 472
metres above cliffs, but the highest summit is that of Slieve League
itself (595 m). Scree slopes descend from the summit towards the sea
and then there are formidable cliffs. A small road continues past
Carrigan Head; then the footpath towards Bunglass gives splendid
views of the cliff face.

Glencolumbkille at the western side of the peninsula is one of
various localities where a remarkable boulder bed is seen. It is an
ancient glacial deposit from high in the Precambrian. Derived pebbles
and boulders of many kinds but especially quartzite, granite, and
dolomite were carried by ice and dropped into the sea. There are different units whose total thickness may exceed 100 metres. The same
deposit may be traced from Mayo through Donegal and Scotland, to
Scandinavia. It provides a reference point representing a particular
time.

The road northwards from Killybegs takes us to Ardara where the
routes diverge around a granite intrusion which postdates the regional
structure and metamorphism. The Ardara Pluton has a roughly circular form about 8 km in diameter. The intrusion has deflected the
structures of the rocks around it. This granite body is earlier than the
Main Donegal Granite, which, near Glenties, cuts across a north-easterly stalk of it. North of Dunglow the highly indented area of the
Rosses is really an old erosion platform about 60 metres above sea
level, scattered with small lakes and bog. Here is another granite,
which, in contrast to the Adara Pluton, came into place by subsidence
(cauldron subsidence) of molten rock at depth which allowed the
magma to occupy a roughly circular area about 8.5 km in diameter, the
outside of which slopes away from the centre. There are several indi-

vidual concentric units. The Adara Pluton was emplaced within the larger Thorr Pluton. The coast road from Bunbeg follows this northwards to Bloody Foreland, the north-westernmost point of Ireland, so called because of its colour in the setting sun. Tory Island and Inishbofin both take in the margin of this pluton. The Thorr granite is coarse grained but varies from a homogeneous type in the north to quartz diorite in the south with many inclusions of surrounding rocks. It is the oldest of the Donegal granites. The origin of this pluton has posed problems. Evidently its intrusion was accompanied by the assimilation of blocks of the surrounding and originally overlying rocks.

This leaves in west Donegal the large area of the Main Donegal Granite, running from near Glenties north-eastwards. Unlike the weak topographical expression of the other granites, this forms the Derryveagh and Glendowan mountains. They are separated by a fault trending from north-east to south-west along which the granite is shattered. It is expressed by river valleys, U-shaped through earlier glacial erosion, and by linear lakes. The fault valley is beautifully displayed in Glenveagh National Park (Fig. 60), with its castle and gardens situated on the eastern side and with views over the lake and wooded slopes.

The Main Donegal Granite is characterised by a strong foliation parallel with its own orientation. It has an aptly named 'ghost stratigraphy' in which trains of other rocks (other granites and Dalradian metamorphic rocks) follow the same trend. Its origin has caused much debate. It seems that all the Donegal granites were emplaced in a tectonically stressed environment. The Main Granite appears to have been intruded as a series of sheets which were sufficiently crystallized as to take up the stress of its surroundings.

A widening Dalradian outcrop lies between the two prongs of the Thorr and Main Donegal granites. There is fine scenery here with a series of quartzite mountains, of which Donegal's highest summit, the immaculate Errigal (2466 m) (Fig. 61), rising from the blanket bog, and Muckish are the best known. Folding in Donegal is on many scales representing different phases. Something of its larger effects are seen in these mountains. The valley occupied by Altan Lough to the north of

Fig. 60: Glenveagh National Park, County Donegal.

Errigal and the gap followed by the road to the south-west of Muckish are examples of the glacial breaching of the original watershed.

Along the north coast of Donegal there is much to be seen. North of Muckish is Horn Head, which, with its attendant small peninsula, is connected to the mainland only by a tombola of blown sand. Here are quartzites, metamorphosed dolerites, and schists. There are steep cliffs exposed to the full force of the Atlantic. To the east of Horn Head is Sheep Haven, with, on its eastern side, another small peninsula tied by blown sand. Here begins the outcrop along the north coast of the Fanad Pluton which extends across the outer part of the intricate Mulroy Bay to Fanad Head and then across the fjord of Lough Swilly. There is likely to be much more of it under the sea. The intrusion of this granodiorite took in various portions of its original roof and surrounding rocks.

Malin Head, the northernmost point of Ireland, is of quartzite. Above the present coast a pre-glacial cliff can be identified and below it a raised beach (Fig. 62). Ten kilometres to the north-east of the head-

Fig. 61: Errigal, County Donegal.

Fig. 62: Old cliff line and raised beach, Malin Head, County Donegal.

Fig. 63: Blue Stack Mountains, County Donegal.

land is the small rocky island of Inishtrahull, of importance in Irish geology. Here are Precambrian gneisses of the same age as the upper division of the Lewisian rocks of north-west Scotland, that is much older than the Dalradian. It appears that the important Great Glen Fault of the Scottish mainland must pass between Inistrahull and the Donegal mainland.

The importance of north-east to south-west trending faults in Donegal has been mentioned already. The most significant is the Leannan Fault. Here a lateral component of movement was particularly important. Faults with significant displacement are frequently the more difficult to trace. In this case the fault is known to cross the southern end of Malin Head, to follow the valley of the Leannan River, and eventually to disappear under the Carboniferous rocks of the Donegal syncline. To the south-east of it the structural trend of the Dalradian rocks changes; quartzites are much less conspicuous; and, with one exception, there are no more granites. The round topped somewhat subdued Sperrin Mountains in counties Derry and Tyrone are made of a variety of metamorphic rocks including schists, thin quartzites, and limestones. Their structures are complex, involving overfolding and hence inversion.

The additional granite is that of the Barnesmore Pluton, which is cut by faults, one crossing Lough Belshade. It forms the remote Blue Stack Mountains of south-west Donegal (Fig. 63). This granite is unusual in not interfering structurally with the schists around it (though there is some thermal metamorphism) and in showing no clear orientation within. The Blue Stack Mountains were crossed by the ice divide. The original watershed has been breached by ice, in particular in the Barnesmore gap followed by the Donegal Town to Letterkenny road. The river here is clearly inadequate to have created this valley. The granite walls were scraped by the ice. The granite is cut by many (much younger) Tertiary dykes, characteristically north-west to south-east in orientation.

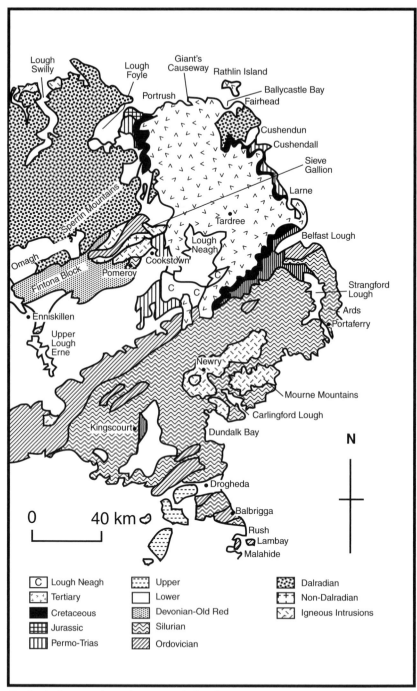

Lough Swilly

Lough Foyle

Giant's Causeway

Rathlin Island

Portrush

Ballycastle Bay

Fairhead

Cushendun

Cushendall

Sperrin Mountains

Sieve Gallion

Larne

Omagh

Tardree

Belfast Lough

Fintona Block

Cookstown

Pomeroy

Lough Neagh

Enniskillen

C

C

Strangford Lough

Ards

Portaferry

Upper Lough Erne

Newry

Mourne Mountains

Kingscourt

Carlingford Lough

Dundalk Bay

N

Drogheda

0   40 km

Balbrigga

Rush

Lambay

Malahide

| C | Lough Neagh |
| Tertiary |
| Cretaceous |
| Jurassic |
| Permo-Trias |

| Upper |
| Lower |
| Devonian-Old Red |
| Silurian |
| Ordovician |

| Dalradian |
| Non-Dalradian |
| Igneous Intrusions |

Geological map of north-eastern Ireland

6

East County Derry, east Tyrone, Antrim

The Dalradian of the north coast of County Donegal ends eastwards at Lough Foyle, as does the county itself. This inlet has a narrow irregular outcrop of Carboniferous clastic rocks round its southern end. Its eastern shore turns inwards until the entrance at the flat Magilligan Point is very narrow. We are now in a very different geological world, with younger rocks than any we have seen earlier except for that very small occurrence of Cretaceous chalk near Killarney. Here is a varied succession of Mesozoic and Tertiary rocks, of which the most conspicuous are the basalt lavas of the Antrim plateau. The structures are much simpler than those in Donegal.

The road from the east side of the River Foyle in Derry City runs north-eastwards past the two small Enagh Loughs. These are kettle holes, where large blocks of ice remaining in the local glacial outwash deposits thawed. Magilligan Point is of blown sand above alternations of shingle and peat, and at depth Triassic and Lower Jurassic rocks. Mudstones and limestones with Jurassic molluscs may be seen in the stream above Tircrevin Bridge. There is a west facing scarp to the Tertiary basalts which meanders southwards, overlooking the valley of the River Roe. The lavas reach a height of 1260 metres in Binevenagh, from which the point can be viewed. To the south-east of the main road as it turns towards the north coast the effects of landslip below the basalt scarp are well seen. The Tertiary basalts and Cretaceous chalk below them have slipped off the Jurassic and Triassic clays below. The basal part of the Antrim chalk has a development of 'greensand' (a glauconitic calcareous sandstone), from which water emerged to lubricate the surface of the slope. The main effect was at the end of the glaciation when there was much water from the retreating ice,

It is possible now to use a coastal, or near coastal, road all the way to Larne in south-east County Antrim. For much of the way the scenery is both rewarding and informative. The plateau has a veritable network of roads. We shall take a clockwise route, as has been the habit, following the coast and then moving inland.

Once blocked by an advance of ice from Scotland, the River Bann now drains Lough Neagh northwards again through Coleraine, built on the wide expanse of what is called the Upper Basalt Formation. There are intrusive as well as extrusive rocks in Antrim; thus at Portrush a sill of dolerite forms Ramore Head along which the town is built. It also makes the small islands of the Skerries offshore. This olivine dolerite sill was first discovered at the end of the 18th century. As can be seen on the foreshore, it has baked the adjacent Lower Jurassic grey mudstones into a hard dark hornfels similar in appearance to the volcanic rock, but in which ammonites (Fig. 64) can still be seen (but on no account extracted). Famously, on the belief that the fossils were actually in dolerite, it was used as evidence for the Neptunist theory that both sedimentary and volcanic rocks had originally been formed in the same way by deposition from a primaeval ocean. In the first years of the new century the mistake was recognised and the views of the rival Vulcanists thus vindicated.

In the stretch of coast from Portrush to Ballycastle, which includes the World Heritage site of the Giant's Causeway (Fig. 7), a fuller sequence of basalt lavas is seen. A Lower Basalt Formation is separated from the more widespread Upper Basalt Formation by an Interbasaltic Formation, in which there are two soils separated by siliceous basalts. Intrusive sills can produce thermal metamorphic effects both below and above, whereas lavas extruded at the surface heat only what is below. These basaltic formations consist of various individual flows, usually some 4 to 6 metres thick, and, in some cases, there was sufficient time between flows to allow the development of a lateritic soil (Fig. 65). Here, after weathering and a final leaching even of silica,

Fig. 64: Baked shale with Lower Jurassic ammonites, Portrush County Antrim.

oxides and hydroxides of iron and aluminium would remain. This is characteristic of tropical regimes, but here the fossil plants from these soils do not suggest such a hot climate; so higher ground water temperatures must have been responsible. The best known feature of the Giant's Causeway is the display of polygonal columns of lava (Fig. 66), produced by cooling and contraction around regularly placed centres.

Farther east, in White Park Bay both Lower Jurassic mudstones and the unconformably overlying White Limestone Formation are seen. The latter is the equivalent of the Upper Cretaceous Chalk of the south of England, though the Antrim rock is more resistant (Fig. 67). At the east end of the bay Lower Jurassic fossils such as ammonites may be found. The chalk contains plates of a characteristic crinoid.

At Ballintoy Harbour there is a fault between columnar basalts and the chalk. The small island of Carrickarade beyond Ballintoy is part of a volcanic vent separated by a chasm from the remainder on the main-

Fig. 65: Red lateritic soil at weathered top of a Tertiary basalt lava flow, County Antrim coast; another lave has followed.

land. Blocks of basalt are seen in a matrix of basalt, chalk, and Jurassic mudstone. A line of chalk cliffs is seen in the bay to the west. The chalk typically has layers of flint nodules of various sizes. These are of cryptocrystalline silica which grew three dimensionally in situ in the original soft sediment and thus developed their sometimes remarkable shapes, reminiscent of some of the sculptures of Henry Moore. Much of the silica may have come from the little rods of silica which build the skeletons of many sponges.

Particularly striking are places where the rock succession or faulting brings the White Limestone in contrasting juxtaposition with the dark basalt lavas. The effect can be seen across the sound in Rathlin Island (Fig. 68). At its western end, for example, the dark lavas can be seen to overlie, and indeed to protect, the chalk.

On the east side of Ballycastle Bay something very different appears on the foreshore: an inlier of Lower Carboniferous rocks. The succession is a varied one: there are sandstones with fossil plants, mudstones, thin ironstones, thin lava flows, and tuffs. The middle of the succession, well seen on the shore is of thin coals (once exploited) and fireclays interbedded with dark shales, some of which are marine. Higher there is a mixture of continental red sandstones and fossiliferous marine limestones. The fireclays are silicified fossil soils in which are the roots of some of the plants that made the typical coal measure swamps where the coal seams originated. The Carboniferous rocks here are cut by various Tertiary dolerite dykes which may be seen to have baked the adjacent rocks. The prominent North Star Dyke is over 3 metres wide. It stands up as a wall of rock crossing the shore slightly west of north. Just to the west of it a red calcareous bed is rich in fossil brachiopods.

Above the Carboniferous, the Fair Head Sill of dolerite, with its columnar jointing, makes an imposing feature in the landscape. It dips slightly southwards; the dip slope can be seen when viewed on the plateau above. It is what is known as a transgressive sill in that it cuts upwards to the east from the Carboniferous, through the Trias, and eventually into the Cretaceous above At the east end of Murlough Bay the Carboniferous is still there on the shore. In the cliff above there are red Triassic sandstones which are followed directly by the Cretaceous. Then we enter yet another example of the remarkable variety of Antrim geology. Here is an inlier of Dalradian rocks faulted against the Carboniferous. Various schists are seen between here and Cushendun, but the small projection of Torr Head is in a limestone formation. Immediately inland from Murlough Bay and Torr Head, an outlier of the White Limestone and Lower Basalt Formation cuts across the different components of the Dalradian sequence.

At Cushendun a south-westerly trending belt of Old Red Sandstone appears, unconformable on the Dalradian. The village itself is built on a raised beach. The succession begins with conglomerate with large rounded and cracked pebbles of quartzite. Then there are cross-stratified red sandstones. The continental sequence is concluded with more conglomerates, becoming increasingly dominated by boulders of andesite, and by interbedded tuffs with andesite boulders, well seen on

Fig. 66: Polygonal columns in Tertiary basalt lava, Giant's Causeway.

Fig. 67: 'Elephant Rock', Sliddery Cove near Portrush, County Antrim. Marine erosion in chalk with rows of flint nodules.

the coast at Salmon Rock. These may relate to a volcanic plug in Cushendal. There is a similar intrusive sheet of quartz andesite to the south of the local river. A further area of red rocks, seen farther south is believed to be of Trias rather than Old Red Sandstone. It is seen on both sides of Glenariff, one of the famed Glens of Antrim, which opens on to Red Bay. Chalk and then basalt form the valley sides, with waterfalls in places.

The splendid Antrim coast road from here to Larne is built close above the sea with steep cliffs above it. Knockore (360 m) south-west of Garron Point affords views across to Scotland. Along here we see the now familiar arrangement of the basalt lavas protecting to a varying extent the Triassic, Jurassic, and Cretaceous rocks beneath. There are examples of landslipping from the cliffs. The substantial inlet of Belfast Lough forms the southern limit of County Antrim. The basalt scarp continues, but now south-westwards. There is a relatively wide area of Trias below, out of which the lough has been cut. A

Fig. 68: Chalk on Antrim coast and Rathlin Island in the distance

swarm of Tertiary dolerite dykes cuts the coast with a west of north orientation.

On the extensive basalt plateau itself there is blanket bog country and a lowland belt of drumlins extends northwards from Lough Neagh. The plateau surface is interrupted in places by small features made by dolerite intrusions or occurrences of acid volcanic rocks. Slemish Mountain, centrally situated, for example, is such a dolerite. At Tardree some 10 km to the south, and 8 km north-west of the north-east corner of Lough Neagh, there is a central dome some 60 m high of columnar jointed rocks surrounded by acid lavas and tuffs. These rhyolitic rocks were extruded onto the weathered surface of the Lower Basalt Formation.

The Lough Neagh Basin had been an area of subsidence ever since Carboniferous times and the Tertiary (Palaeocene) lavas are here at their thickest. Boreholes have given a maximum figure of over 780 metres. But this subsidence continued after the volcanic activity had ceased. Over 350 metres of mostly lacustrine and swamp deposits, the so called Lough Neagh Clays, overlie the basalts. They are Oligocene in age, this being shown by palynomorphs from what evidently was a lowland swamp vegetation. The shallow Lough Neagh is the largest body of freshwater in the British Isles, but perhaps less conspicuous because it is not surrounded by dramatic country. Its earlier and present drainage were mentioned previously. To the south-west the clays are exposed on the land They cut across the edge of the basalt lavas here and this south-east corner is the only place where the familiar scarp is not present. That the subsiding basin is not confined in effect to Lough Neagh is shown by the basalts as a whole, which slope inwards towards the centre of the whole plateau from both east and west.

Awkwardly placed geographically in relation to our coverage of other regions, but best considered here, is that approximately rectangular belt of country between Magherafelt and Enniskillen. It is separated from the Dalradian to the south of the Sperrin Mountains by the south-westerly trending Omagh Fault.

To the north, north-east, and south-west of Cookstown, first of all, is a relatively small area of varied and complex geology. The core of it is the so-called Central Inlier of Tyrone of highly metamorphosed sedimentary rocks and granitic pegmatites. These rocks may, as was the case with some in the Ox Mountains, belong to the very oldest part of the Dalradian. Overlying these and cropping out at both sides are rocks of the Tyrone Igneous Complex. Here a 'basic plutonic complex' is associated with a varied group of volcanic rocks, followed by the intrusion of small granitic bodies. One of these forms Slieve Gallion (527 m) at the north-eastern end of the complex. Black shales associated with volcanic rocks exposed on the side of this mountain have yielded graptolites of Early Ordovician age. Finally, to the south-east of the igneous complex, a small faulted and folded area close to Pomeroy has long been known for its wealth of Lower Palaeozoic fossils. There are late Ordovician shallow water deposits with brachiopods and trilobites and early Silurian richly graptolitic shales.

The other part of the belt of country under consideration forms a subsidiary rectangle north-east of Lower Lough Erne and Enniskillen. This is the 'Fintona Block' of undulating country rising locally to 305 metres. The poorly exposed rocks were once though to be all Old Red Sandstone conglomerates and sandstones, derived variously; but palynology (study of pollen and spores) has shown that nearly a third of the area, extending as a faulted triangle north-eastwards from a base near Enniskillen is actually of Carboniferous sandstones.

The south-eastern fault of the converging pair which delimit the Carboniferous triangle at its end at Lesbellaw contributes to the presence of a very small inlier of Silurian rocks. Here slumped conglomerates with clasts derived from the Dalradian and the Tyrone Igneous Complex are sandwiched within graptolitic shales of Llandovery age. The village of Lisbellaw itself is built upon one of the features made by the conglomerates in which some of the clasts reach 0.6 m in diameter.

7

Down, Armagh, Monaghan, Cavan, Longford, and Louth

From the east of Belfast the outcrop of Triassic rocks curves south-eastwards past Newtownards into Strangford Lough, its name derived from the Danish 'strang fjord' after the swift tidal currents at its narrow entrance near Portaferry. On the hill of Scrabo near Newtownards a Tertiary dolerite sill forms a capping. It has cut into and bleached the red sandstones of the Trias. Otherwise, from the east coast of the Ards Peninsula, south-westwards across country as far as County Longford and southwards almost to Drogheda, there is an extensive outcrop of folded and variously cleaved Lower Palaeozoic (Ordovician and Silurian) deep water rocks, often referred to as the Longford Down massif. It is an obvious continuation, geologically speaking, of the Southern Uplands of Scotland, though the topography in Ireland is relatively subdued.

Within the Longford Down massif, a narrow belt of Ordovician rocks is followed to the south-east by a wider tract dominated by the Silurian, but with faulted strips of Ordovician. Within the northernmost part of the Ordovician outcrop there are greywackes with some finer-grained rocks and volcanics. These are separated by a most important fault, the Orlock Bridge Fault, from a belt of graptolitic shales to the south. The Silurian rocks again are of greywackes and graptolitic shales, the former on the whole having arrived later in the succession southwards. The Longford Down massif is crossed by a number of strike faults; within each faulted strip the beds tend to dip northwards, but the sequence as a whole becomes younger southwards.

The graptolite faunas in the Ards Peninsula, both inland and along the east coast, have long been known for their full and detailed succession.

In the massif as a whole, but particularly farther east, there is some topographical expression where more resistant greywackes or quartzitic grain flows form crags. In contrast, the narrow faulted strips of Ordovician shales within the Silurian outcrop form hollows with lakes in places. However, the most obvious scenic feature everywhere is of the great belt of drumlins, which may be seen as ending in small islands on the east coast of Strangford Lough. Inland they produce a topography with the hollows between them occupied by small lakes or bog, and within which one may feel cut off from the world outside.

An outlier of younger rocks, orientated from north to south is conspicuous within the Lower Palaeozoic outcrop near Kingscourt (County Cavan). The succession rises towards its western faulted margin. The Lower and Upper Carboniferous, Permian, and Trias are all represented. The Permian includes economically exploited gypsum deposits. The Permo-Trias here is an isolated remnant of the more extensive record from boreholes and surface outcrops in Antrim of the arid coastal continental and saline environments of the time.

In contrast to all the above and dominant in the scenery is the south-eastern mountainous area which surrounds Carlingford Lough. The lough is an over-deepened glacial valley with evidence of moraines sweeping along its north-eastern side and curving southwards. An isolated low-lying area, floored by Carboniferous rocks at its seaward end, and especially to the south, forms the so-called Kingdom of Mourne. Inland from the lough and following the north-easterly Caledonoid trend for some 40 km through Newry is the relatively large area of the Newry Igneous Complex. Inconspicuous in the topography it includes along its length three separate Caledonian plutons of granodiorite.

The mountains are made of three groups of much younger Tertiary rocks. Intruded into the south-western end of the Newry Igneous Complex is that of Slieve Gullion (583 m), probably representing the core of a volcano. Slieve Gullion itself shows successive intruded layers of porphyritic rock and dolerite. Surrounding it is lower ground on the

Fig. 69: Carlingford Lough.

Newry Granite and then, outside that, an obvious ring of hills some 10 km in diameter produced by porphyritic ring dykes related to a circular fracture produced by subsidence.

Farther south-east along the peninsula which separates Carlingford Lough (Fig. 69) and Dundalk Bay is the intrusive centre of the Carlingford Mountains. This cuts both Silurian and Carboniferous rocks. Again the geological history is complex; essentially there is a centre of porphyritic rock which is surrounded discontinuously by gabbro, which makes the high ground. Glenmore cuts into the centre but its orientation is also fault controlled.

Fig. 70: Newcastle with Slieve Donard, Mourne Mountains.

This brings us to the third and largest of the three Tertiary igneous centres, that of the Mourne Mountains (Fig. 70). Their highest peak of Slieve Donard reaches over 850 metres. A distinction can be made between the eastern or High Mournes, with their more dramatic topography, and the western or Low Mournes. Both are of granites but their division is a clear topographical break with penetration from the north of a tongue of Silurian rocks. We are here seeing a deeper level of intrusive activity. Three separate granites are involved in the eastern area and two in the west, the former in general more porphyritic. Cauldron subsidence was involved in the intrusion of the succession of granites.

Fig. 71: The Silent Valley, Eastern Mournes.

The surrounding Silurian rocks have been thermally metamorphosed and thus hardened to form hills. The coastal road to the east however shows the granite steeply plunging The high summits of the Mournes remained above the Quaternary ice but corries are well developed. The glaciated valleys of the Annalong River and the Silent Valley (Fig. 71) run generally southwards through the eastern Mournes.

8

Drogheda to Howth

The last part of our journey through Irish scenery can be taken from Drogheda on the River Boyne southwards to the Howth Peninsula, where we started. Up the Boyne valley from Drogheda is the splendid archaeological site of Newgrange. Towards the coast the river has cut a gorge in Carboniferous limestone, which we have now reached beyond the southern limit of the Lower Palaeozoic rocks of the Longford-Down massif. Farther south the Lower Palaeozoic rocks are encountered again in an inlier which reaches the coast around Balbriggan. The solid rocks are conspicuous only on the coast. Close to Balbriggan, Ordovician submarine volcanic rocks of andesitic composition are seen as pillow lavas, breccias, and tuffs. They are interbedded with, and succeeded by, graptolitic mudstones. From about one kilometre south of the town and for a further two kilometres along the coast is perhaps the best Silurian graptolitic sequence (Llandovery and Wenlock) in the whole of Ireland. Some 50 species have been recorded. These rocks, even to the extent of some banded bluish mudstones in the Wenlock, are very similar to those of the same age in the English Lake District. High in the Wenlock, a formation of greywackes with only fragmentary graptolites appears.

The Carboniferous is found again to the south of Skerries. The coastal section shows various sedimentary and tectonic features of these rocks. There are most impressive Variscan folds at Loughshinny (Fig. 72). At Rush Harbour, the youngest beds are oolitic limestones seen on the north side. On the south side there are conglomerates. Farther south again some turbiditic sandy limestones and a remarkable

Fig. 72: Folded Lower Carboniferous limestones and mudstones, Loughshinny, County Dublin.

conglomerate with big boulders are present in the upper part of the so-called Rush Slates. High in the slates abundant goniatites can be collected.

Lower Palaeozoic rocks appear again in Lambay Island and on the mainland at Portrane. The Ordovician on Lambay, as on the mainland, is present mainly as andesitic lavas and pyroclastic rocks, but includes also fossiliferous limestones and mudstones. A very distinctive rock here is the so-called Lambay Porphyry. A more accessible section is found on the mainland, where there is much of interest. It is best to begin by the minor road, which, having reached Portrane from Donabate, continues to the coastal car park by a martello tower. Southwards then there is a cliff path with access to the section along the shore. A conglomerate of basal Old Red Sandstone type is seen near the road, unconformable on the Ordovician andesitic volcanics. There are minor intrusions of Lambay Porphyry. Associated with the volcanic rocks are thin limestones with trilobites and black graptolite bearing mudstones. The volcanics appear to have come from an oceanic volcano situated between what are now Lambay and the coast,

Fig. 73: Silurian rocks thrust over Ordovician, Priest's Chamber, Portrane.

with the sediments forming against it. Younger than these, and faulted against them, is the highly fossiliferous Portrane Limestone. A rich silicified fauna of brachiopods, corals, bryzoans, etc. allows precise dating to the top series of the Ordovician. There is evidence of soft sediment deformation in the thinner bedded limestones. Part of the coastal section is a more massive reef limestone some 10 metres thick. In Priest's Chamber (Fig. 73), Silurian rocks are seen to be thrust northwards over the Ordovician. These are well-bedded greywacke sandstones, siltstones, and mudstones which resemble Silurian rocks elsewhere. The southern end of the narrow outcrop may show an unconformable relationship with the underlying Ordovician.

We come once again to well exposed Carboniferous rocks just below the coastal road to the south-east of Malahide. Extensive bedding planes are seen in the north. Farther south, the section is crossed by various folds and faults. The limestones vary in their detailed lithology. There are good coral faunas. The main road now continues through Portmarnock, Baldoyle, and Sutton to the Howth peninsula where we began our journey.

Part 4

THE GEOLOGICAL
BACKGROUND

I

The Rocks beneath our Feet

Our tour through Irish scenery has shown how much depends upon the rocks beneath our feet; the scenery we see, that is to say, depends upon the materials of which the Earth is made. If we look at a granite in the Wicklow Mountains (Fig. 15), or in a Dublin building where it has been employed as a building stone, we see that the rock itself is an aggregate of things smaller and simpler. These smaller components are **minerals**. A mineral is a naturally occurring inorganic substance characterized by a chemical composition and physical properties that are either constant or variable within a definite restricted range. These depend upon the chemical elements of which the mineral is made and on the way they are arranged. There are some 2000 known minerals, but, fortunately, of these only about 15 to 20 make up the bulk of the rocks.

The mineral most commonly recognised is quartz, be it as the glassy looking component of granite, as the commonest grains of sand or sandstone, as resistant white veins, or occasionally as beautifully formed crystals. These last display an obvious geometrical shape of a prism with pyramids. In the science of crystallography various classes of such geometry are recognised, they themselves depending upon the molecular arrangements within the crystal. For another example, mica occurs in its familiar shiny sheets because the molecular components within it are themselves arranged in sheets. The basic architectural units are tetrahedra in which 4 oxygen atoms are packed round a silicon atom. Again, in rock salt, which has a cubic geometry, the sodium and chlorine atoms within are so arranged that 6 sodiums surround each chlorine and 6 chlorines surround each sodium, with all these orientations at 90 degrees to produce a cuboidal pattern.

To identify common minerals such as those mentioned we do not need microscopy or such more sophisticated techniques as electron microscopy, important though these are; we can recognise them by simpler observations, not only of crystal form but also of colour, lustre, hardness, specific gravity, cleavage (which refers to the tendency as in mica for minerals to split along planes), mode of fracture, or even in special cases of taste or magnetism.

In terms of composition, the most frequently found minerals are silicates. In quartz, referred to already, the silicon is combined with oxygen to give the formula SiO_2. Taking the matter of composition in a geochemical way, it may seem surprising that more than three quarters by weight of the Earth's crust is of oxygen and silicon, the former making 47 percent by weight and 92 percent by volume. Quartz is a hard resistant mineral, which explains its frequent presence in sand.

The other especially common minerals in granite (though they occur in many other rocks) are the feldspars, which are of silica and oxygen along with potassium, aluminium, sodium, and calcium. Orthoclase feldspar is the potassium aluminium silicate. The plagioclase feldspars have a range of composition according to the proportions of sodium and calcium that they contain. The feldspars are commonly white or pink in colour. They are more easily weathered than is quartz. The micas are also aluminous silicates with various other elements. They make a group with the so-called clay mineral which form the very fine grained mudstones or shales.

A second great group of rock-formers are the ferro-magnesian minerals. In all these silicon and oxygen are combined with iron and magnesium in different structural arrangements. They are dark coloured minerals, black, brownish, or green, typically found in rocks like basalt. They include such minerals as augite, olivine, and hornblende.

Apart from the silicates, another important group comprises the carbonates, particularly calcite, the familiar calcium carbonate of the limestones. The oxide minerals include the various iron oxides, of which hematite is responsible for the colour of our common red sandstones. Rock salt (halite) mentioned above is a chloride. Other economically important minerals are gypsum, which is a calcium sulphate, and the relatively heavy barium sulphate known as barite. Such minerals as these may be present as substantial deposits. Lastly, the various sulphide minerals present as veins or disseminations are nevertheless economically significant. They include the grey lead sulphide galena; the more brownish zinc sulphide sphalerite; and pyrite, the golden yellow iron sulphide, which is popularly known as fool's gold.

Rocks have their individual mineral compositions and underlying chemistry. They are characterised also by a second basic property, that of texture. Texture is the inter-relationship of the various components of a rock, including their sizes and shapes. Rocks can be placed in three traditional categories: igneous, sedimentary, and metamorphic.

The **igneous rocks** are formed by processes of internal origin in the Earth. They may themselves be subdivided into *intrusive* and *extrusive* types. The intrusive igneous rocks crystallised at depth, more or less slowly, from a hot melt referred to as a magma. They are exposed at the present surface only after erosion has removed the once over-lying rocks. The extrusive or volcanic rocks, on the other hand, have solidified from lavas poured out at the Earth's surface, where they have cooled relatively quickly; or they may have accumulated as fragments emitted explosively from volcanoes. The latter are the pyroclastic rocks (literally broken by fire).

It is possible to go some way towards classifying igneous rocks simply on the basis of their mineral composition (Fig. 74). A simple table (Fig. 75) shows some of the most common igneous rock types in relation to this classification, but here an element of texture has been added. The quartz porphyry shown on the table is a very striking rock type, known even to the Romans, in which large crystals (phenocrysts) of quartz and sometimes feldspar are set in a finer grained background. It is understood that the large crystals grew slowly at first in the liquid melt at depth and then, as the intrusion continued to rise, cooling accelerated and the finer matrix formed. Pegmatites (not shown) are

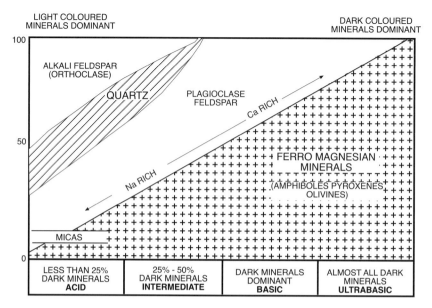

Fig. 74: Classification of igneous rocks based on mineral composition.

exceptionally coarse grained rocks representing late stages of crystallization.

Early in the history of the geological sciences it became recognised that ancient volcanic rocks must be the results of the same processes that can be seen in operation in volcanic regions at the present time; direct comparisons are possible. With the intrusive igneous rocks it is different. They were formed originally at depth and all we can do is to assess examples from the past. Common small-scale, or minor, intrusions are the dykes and sills, the former, to put it at its simplest, are relatively narrow sheets of igneous rock that cut upwards across the layering of the surrounding rocks, while the latter behave on the whole concordantly along the layering. Other minor intrusions are the volcanic plugs which rise through previous vents and solidify there. Major igneous intrusions (plutons) range in size up to tens or even hundreds of kilometres in the case of the vast batholiths of the western United States, their magnitude bringing problems of interpretation.

Perhaps the most striking feature of the **sedimentary rocks** is that they are usually stratified; that is they occur as more or less distinct,

	COARSE GRAINED	MEDIUM GRAINED	FINE GRAINED	
ACID	Granite	Microgranite	Quartz Porphyry	Rhyolite
	Granodiorite			
INTERMEDIATE	Diorite Syenite		Andesite Trachyte	
BASIC	Gabbro	Dolerite	Basalt	
ULTRABASIC	Peridotite			

Fig. 75: Common igneous rock types.

roughly parallel, layers or beds. The differences expressed in stratification are those of composition, colour, and texture, which represent original changes in formation.

Many sedimentary rocks have a clastic (or broken up) texture, produced as the individual particles (grains or clasts) were deposited from water, wind, or even by gravity alone. Particle size is critical in their classification, with boulder or pebble sized components in conglomerates, sand sized particles in the sandstones, and clay sized particles forming the mudstones and shales. The shape of the particles, too, may be characteristic. Thus the boulders in a stream tend to be more or less well rounded; the material in a scree is angular (the equivalent solid rocks respectively are conglomerates and breccias); grains deposited by the wind tend to be very well rounded. The sediments originally formed are, of course, gradually compacted as more material is added, or are subjected to subsequent physical and chemical

Fig. 76: Flute moulds on inverted Silurian greywacke, near Galloways Burn, County Down. Eddies excavated in the muddy sea floor were filled with sediment as the strength of the turbidity current subsided. They are preserved as moulds on the underside of the greywacke.

changes, known collectively as diagenesis, to produce the sedimentary rocks as we now see them

The rocks of sand size grade in particular show important variations from what might be called a 'normal sandstone' with dominant quartz grains and some other components. The quartzites are exceptionally pure sandstones in which the quartz grains themselves may be cemented by more silica. The common greywackes (an old German term) are ill-sorted sandstones in which the quartz is accompanied by fragments of earlier rock and set in a kind of paste of clay. They are usually the result of turbid flows of sediment down a bottom slope. They may exhibit what is called graded bedding in which the coarser material is found at the bottom of each layer, this having settled first. Bottom structures such as flute moulds (Fig. 76) also are characteristic of such turbid conditions. The foregoing are examples of primary structures in sedimentary rocks. Another such is cross-stratification (Fig. 30), where material deposited from a current in water or air may be stacked in oblique layers at an angle to the more general stratification.

Secondly, there is another group of sedimentary rocks, less common than the above, in which the texture is, at least partially, crystalline or cryptocrystalline ('hidden crystalline') rather then clastic. The siliceous representatives here are the cherts or flints. The former may be bedded or may form discrete nodular masses (nodules, concretions) within the surrounding rock. The term flint is used particularly for such nodular masses in the Chalk. The so-called salt deposits or evaporites include rock salt, gypsum, and anhydrite (the anhydrous equivalent of gypsum), which may form substantial bedded sequences, originally produced by evaporation in salt lakes or restricted marine basins.

Thirdly, abundant and often economically important sedimentary rocks are the limestones. Their purity varies but the dominant constituent is calcium carbonate. Less common is dolomite (calcium magnesium carbonate), which tends to occur as a secondary replacement of the original calcite. The limestones form a very variable group. There are representatives such as the oolitic limestones, where very small, concentrically built up spheres have been precipitated around minute nuclei, keeping their shapes as they grow in the manner of a snowball as they are gently wafted about in warm water. Chalk is a special case of an extremely fine-grained limestone largely of organic origin, built by the microscopic skeletons of marine plants. In many limestones there is a proportion, sometimes a dominance, of shelly material. Finally, limestones are especially prone to secondary (diagenetic) processes and frequently are seen as crystalline rocks.

Metamorphic rocks result from heat and pressure acting upon previously existing rocks, producing fundamental changes in their mineral composition and their texture. As a result mineral grains may be deformed or reorientated; they may be recrystallized into larger grains; or new minerals may grow at the expense of the old ones. Examples of characteristic new minerals are chlorite (a green silicate mineral) and

			ROCK NAME	DOMINANT MINERAL COMPOSITION	SOME PARENT ROCKS
TEXTURE	NON-FOLIATED		Quartzite	Quartz	Sandstone
			Marble	Calcite and/or Dolomite	Limestone or Dolomite
	FOLIATED	Very Fine Grained	Slate	Micas, Chlorite, Serpentine, Horneblende, etc.	Mudstone
					Volcanics
		Medium Grained	Schist		Shale, Intermediate and Basic Igneous Rocks
		Medium-Coarse Grained	Gneiss	Feldspar, Quartz, etc.	Granite, Greywacke.

Fig. 77: Common metamorphic rocks.

garnet. In contact metamorphism an igneous intrusion may literally bake the surrounding rocks for a short distance from its margin. Regional metamorphism is a more widespread and pervasive phenomenon in which, as a result of movements within the earth, a new regime of heat and pressure is established. Rocks such as quartzites which are made from just one mineral, quartz, can only change in texture. The same comment would apply to a pure limestone, but in many cases the presence of some impurity allows the possibility for the growth of new minerals to give one of the colourful marbles. The combination of pressure and temperature in metamorphism results in the development of a characteristic layering known as foliation. In what are called gneisses the foliation is formed by segregation of minerals into roughly parallel discontinuous bands. In schists the finer foliation (silky in appearance) relates to the parallel orientation of platy micaceous minerals.

2

Reading the Rocks

Even in Part 1 of this book it was difficult to avoid the use of some names of kinds of rock, granite and sandstone for example. The common ones have been covered in the last chapter. Other names which must appear if any meaningful description is to be given relate to the ages of the rocks beneath our feet; Carboniferous rocks, for example, have profound effects on Irish scenery; they are strikingly obvious as a central splash, conventionally of blue, on a geological map of Ireland.

The ages of rocks concern us also because the sequence of rocks represents a sequence of events in the history of the Earth. This long geological history unfolds for us as we walk the Irish ground. This chapter, then, is intended to explain the geological time-scale that we use and to introduce the remarkable geological history of this island, poised upon the edge of Europe. Our very long history involves various episodes of earth movements as well as a very varied story of deposition, erosion, and igneous activity. Such processes have affected not only the types of rocks we see but also the ways in which they are now arranged. Thus some short account of geological structures also is included here.

The geological time-scale (Fig. 3)

A way of seeing the past preserved in stone was first glimpsed as long ago as the 15th century, but it was a hundred years or so before the idea was more widely developed. Crucial was the recognition that in a sequence of rocks, as it was simply seen, the oldest strata lay at the bottom and the youngest at the top. Stratigraphy is the study of such sequences and their interpretation as sequences of events in the history of the Earth. Successions of rock are examined, described, and recorded locally; but the heart of the matter is the *correlation* (establishment as far as possible of time equivalence) of these, so as to build up a regional, and eventually world wide, picture of the way the Earth has evolved.

The geological time-scale used in stratigraphy has been built up historically. Thus most of the geological systems shown in Figure 3 were named originally in Europe. The name of the Cambrian System, for instance, was taken from the old name for Wales; the Jurassic System was named from the Jura Mountains of France and Switzerland, where these rocks are well seen. The most effective way of achieving correlation with this time-scale is by the use of fossils. As living things have evolved, so fossils vary with age and can be used as indices of different levels. Thus the ammonites (Fig. 64), for example, well known to many of us as especially beautiful fossils, provide remarkably precise indicators of relative Jurassic time, allowing correlation even within some 150,000 years (and this some 175 million years ago). In Ordovician and Silurian rocks graptolites (Fig. 78) are especially useful and can achieve correlation of less than a million years. In the very old Precambrian rocks, once thought incorrectly to be without fossils, the record does become meagre and other methods of relating sequences must be found. In very young rocks, additional methods come into play, involving climatic change, variations in the Earth's magnetic field, etc.

Apart from terms such as the Jurassic System, we may refer to the corresponding Jurassic Period. This is simply to allow sensible language. We cannot say that the dinosaurs lived in the Jurassic System; we must say that they lived in the Jurassic Period. It is of less concern to us here that the Systems are divided into Series and Stages, but we should recognise that all these divisions are being standardized at internationally agreed places.

Alongside the Systems, Figure 3 also shows dates in millions of year, which we need if the rates of processes in the Earth are to be appreciated. Achievement of this time-scale in years has a long, and sometimes

Fig. 78: Two characteristic Silurian (Wenlock) graptolites. Pristiograptus dubius *(on the left) is preserved in the original organic skeletal material.* Monograptus flexilis *(on the right) shows the skeleton replaced by a white mineral. In P.* dubius *the sicula from which the colony grew is clearly seen below the succession of thecae, each of which housed an individual animal (zooid). Magnification approximately x 5.*

quite acrimonious history. The eventual solution came through the discovery of radioactivity. The nuclei of the atoms of the radioactive elements spontaneously and progressively change to more stable forms, emitting alpha particles or capturing electrons. This process may go through a whole series of changes; but, most importantly, these changes take place at known fixed rates, which individually may vary from a fraction of a second to millions of years. Thus radioactive minerals can provide a kind of clock by which, by estimating the extent of their decay, their ages can be calculated. The first method employed in this way involved changes from isotopes of uranium to the end products, lead and helium. Uranium minerals are rare, but now other such series, some appropriate to particular situations, have come into use. Moreover, laboratories about the world have developed for the increasingly precise dating of rocks. Detailed figures may change somewhat from time to time, but what matters is that this time-scale in years is now essentially stable.

A long, long history

Earthquake waves travel in different ways and at different velocities according to the density of the media through which they pass. Knowledge of this, together with what we know of the rocks at the Earth's surface and from deep boreholes, has led to an understanding of the deep layered structure of the Earth. A central *core* of very high density is surrounded by the *mantle*. Above this is the *crust*. Oceanographical studies have contributed to the picture and it is of fundamental importance that oceanic crust is different from that under the continents. The former is of basalt, with but a thin layer of deep sea deposits: commonly, oozes of various kinds. Continental crust is of variable but much greater thickness. Here a granitic layer overlies the basalt and this is succeeded by the very varied and sometimes very thick sedimentary rocks.

In recent decades the idea of *plate tectonics* has come to dominate our world picture. A number of rigid plates, composed of upper mantle material and crust, move across the Earth at rates of the order of a few centimetres per year. They are generated at lines such as the mid-Atlantic ridge, where new basaltic material is emerging, as is seen so well in Iceland, which lies upon the ridge. Correspondingly the plates are lost in the so-called subduction zones where two of them meet. Where a continental plate is dragged down beneath an oceanic plate volcanoes may occur; where two continental plates meet, as in northern India, a range of mountains may be elevated. It is supposed that convection currents (like those one sees, for instance, in boiling jam) in the lower mantle may be responsible for the movement of the plates themselves.

The earlier history of the present plate tectonic regime resulted in the Mesozoic and somewhat earlier rocks of the great southern continent of Gondwana becoming separated into Africa, South America, Australasia, and Antarctica. Similarly, North America drifted away from Europe. The Antrim basalts relate to these movements.

There appears to have been a grand cyclical evolution in earth history. One earlier plate tectonic cycle began in the late Precambrian when an ancient ocean, the so called Iapetus, appeared. Its final closure was in Silurian times. Lower Palaeozoic rocks now seen in the north-west of Ireland relate to those in parts of North America in having formed on the north-western side of Iapetus. Other Irish Lower Palaeozoic rocks belong with those of much of Britain in representing the south-eastern side of the ocean. Other details of all this, and of older and younger rocks, have emerged in our detailed tour of Irish scenery. To this end, the whole geological evolution of Ireland is summarised in Figure 79.

Bends and fractures

The record of earth movements, major and minor, is, as it were, frozen for our examination in the rocks in the form of folds and fractures. There are many technical terms for these so-called *tectonic structures*. The attitude of rocks is referred to in terms of *dip* and *strike* (Fig. 80A). The dip is the maximum angle between the horizontal and a plane of rock; strike is perpendicular to this. There are two main kinds (Fig. 80B) of *folds*: *anticlines* are upward folds, where subsequent erosion results in a pattern on the ground, or on a map, with the older rocks in the middle. *Synclines* are downward folds in which the opposite applies. Folds may plunge (Fig. 80C), that is to say the hinge where the fold turns over may make an angle with the horizontal. If this effect is not confined to one direction a structural dome or basin will have resulted. Folds also may be symmetrical or asymmetrical. In the latter case the dips on the two limbs are unequal, in some cases even to the extent that one limb becomes inverted.

Folding may be accompanied by *cleavage* (not to be confused with cleavage in minerals), in which, because of compression and perhaps

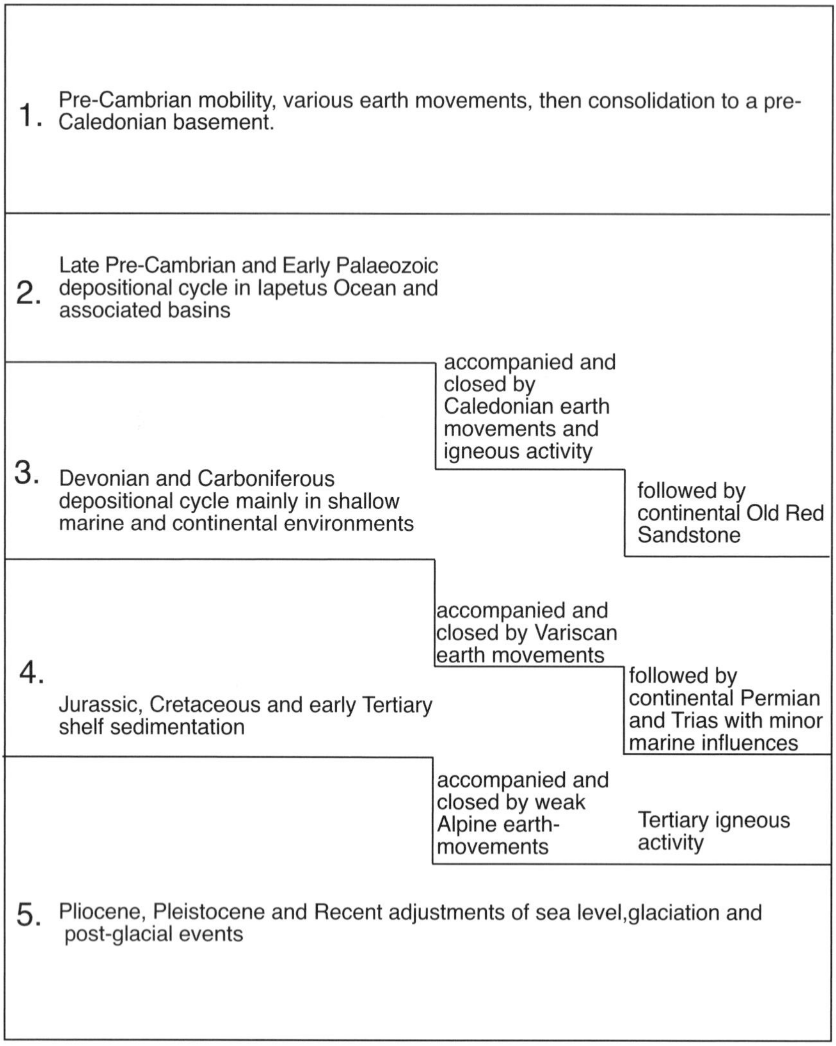

1. Pre-Cambrian mobility, various earth movements, then consolidation to a pre-Caledonian basement.

2. Late Pre-Cambrian and Early Palaeozoic depositional cycle in Iapetus Ocean and associated basins

accompanied and closed by Caledonian earth movements and igneous activity

3. Devonian and Carboniferous depositional cycle mainly in shallow marine and continental environments

followed by continental Old Red Sandstone

accompanied and closed by Variscan earth movements

4.

followed by continental Permian and Trias with minor marine influences

Jurassic, Cretaceous and early Tertiary shelf sedimentation

accompanied and closed by weak Alpine earth-movements

Tertiary igneous activity

5. Pliocene, Pleistocene and Recent adjustments of sea level, glaciation and post-glacial events

Fig. 79: Summary of the geological history of Ireland.

the influence of fluids, rocks separate and fuse along close parallel planes perpendicular to the direction of compression (Fig. 81), thus demonstrating the original stress pattern. Thus these planes are approximately parallel to what we call the axial plane of the fold, the hypothetical plane that bisects the angle between the two limbs.

Folds imply some plasticity of deformation. More rigid effects result

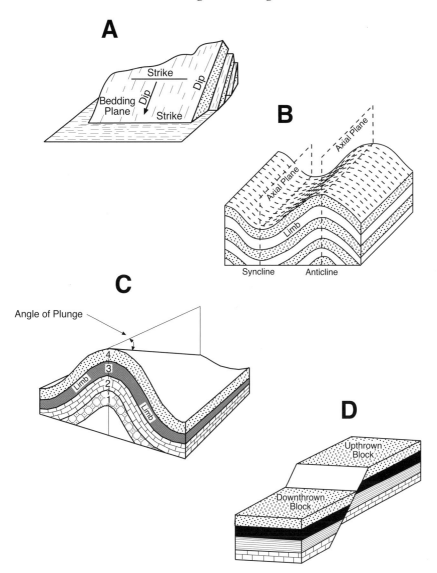

Fig. 80: Tectonic structures and their terminology.

in fractures, either *joints*, which are cracks with no evident movement along them, or *faults* (Fig. 80D), where there is movement, which may be anything from a few millimetres to tens or even hundreds of kilometres. The really large ones occur usually where there is lateral rather than vertical movement between the two sides of the fault plane.

Very common also are examples of what is best referred to as *soft sediment deformation*, sediment having become unstable and sliding down the sea (or lake) floor to produce slump folds.

Fig. 81: Cleavage in Old Red Sandstone rocks. Bedding dips steeply to the right; cleavage steeply to the left.

Illustration credits

The following figures are photographs by the author: 7, 9, 11, 12, 13, 14, 15, 19, 20, 23, 26, 27, 30, 33 and cover illustration, 34, 36, 39, 40, 41, 43, 50, 51, 52, 53, 54, 57, 59, 65, 66, 68, 73, 81.

Other permissions and copyright are gratefully acknowledged as follows:

Aerofilms copyright 25; Bord Failte 8, 16, 24, 38, 47, 48, 58; British Geological Survey IPR/41-9C, Copyright NERC, all rights reserved 64, 67, 76; Cambridge University Collection of Air Photographs copyright reserved 22, 46, 63; Dúchas The Heritage Service copyright 4, 5, 6, 60; Limerick City Gallery (Permanent Collection) 55; Northern Ireland Tourist Board *frontispiece*, 69, 70, 71.

Also: P. Coxon, Trinity College Dublin 49, 61; J. Feehan, University College Dublin 44; J.R. Graham, Trinity College Dublin 37, 56, 72; M.A. Parkes with Permission of the Geological Survey of Ireland 28; M.E. Philcox, Trinity College Dublin 17; M. Quigley, Trinity College Dublin 62; R.B. Rickards, University of Cambridge 78; G.D. Sevastopulo, Trinity College Dublin 35.

Some supplementary reading

Holland, C.H. (Editor). 2001. *The Geology of Ireland*. Dunedin Academic Press, Edinburgh, 531 pp.

Ryan, M. (Editor). 1991. *The Illustrated Archaeology of Ireland*. Town House and Country House, Dublin, 224 pp.

Aalen F.H.A., Whelan, K., and Stout, M. 1997. *Atlas of the Irish Rural Landscape*. Cork University Press, 352 pp.

Index of place names

Note: page numbers in italic denote illustrations.

Index of place names

Selected index of technical terms